The Successful

ELECTRONICS
TECHNICIAN

12 Essential Strategies for Building a Winning Career

The Successful

ELECTRONICS TECHNICIAN

12 Essential Strategies for Building a Winning Career

By
David L. Goetsch

THOMSON

DELMAR LEARNING Australia • Brazil • Canada • Mexico • Singapore • Spain • United Kingdom • United States

THOMSON

DELMAR LEARNING ™

The Successful Electronics Technician:
12 Essential Strategies for Building a Winning Career
David Goetsch

Vice President, Technology and Trades ABU:
David Garza

Director of Learning Solutions:
Sandy Clark

Managing Editor:
Larry Main

Senior Acquisitions Editor:
Stephen Helba

Marketing Director:
Deborah S. Yarnell

Marketing Manager:
Guy Baskaran

Marketing Coordinator:
Shanna Gibbs

Art Director:
Jack Pendleton

Development:
Dawn Daugherty

Director of Production:
Patty Stephan

Production Manager:
Andrew Crouth

Content Project Manager:
Benj Gleeksman

Library of Congress Cataloging-in-Publication Data

Goetsch, David L.
  The successful electronics technician : 12 essential strategies for building a winning career / David Goetsch.
      p. cm.
  Includes index.
  ISBN 1-4180-6176-X
  1. Electronics—Vocational guidance. 2. Electronic technicians. I. Title.
  TK7845.G64 2008
  621.381023—dc22
                                            2007016713

NOTICE TO THE READER

Publisher does not warrant or guarantee any of the products described herein or perform any independent analysis in connection with any of the product information contained herein. Publisher does not assume, and expressly disclaims, any obligation to obtain and include information other than that provided to it by the manufacturer.

The reader is expressly warned to consider and adopt all safety precautions that might be indicated by the activities herein and to avoid all potential hazards. By following the instructions contained herein, the reader willingly assumes all risks in connection with such instructions.

The publisher makes no representation or warranties of any kind, including but not limited to, the warranties of fitness for particular purpose or merchantability, nor are any such representations implied with respect to the material set forth herein, and the publisher takes no responsibility with respect to such material. The publisher shall not be liable for any special, consequential, or exemplary damages resulting, in whole or part, from the readers' use of, or reliance upon, this material.

TABLE OF CONTENTS

INTRODUCTION

The Successful Electronics Technician

Electronics is a broad and varied career field. Consequently, the title *electronics technician* is also broad and varied. It can be applied to a number of specific positions in several different areas of specialization. Electronics technicians may be found working in most of the major industrial sectors, including aviation, automotive, manufacturing, computers, radio, healthcare, and television. They can also be found working in federal, state, and local government positions as well as in the various military services. We live in an electronic world, and the electronics technician keeps that world spinning properly. Therefore, electronics is a high-demand, high wage career field, which is good news for you.

High demand means that there are plenty of good jobs available in the field for those who have the right education and training. High wage means that those jobs pay well. In fact, electronics is one of the highest-paying career fields a person can pursue as a technician. Electronics is also a high-skills career field, which means that not everyone can be an electronics technician. Only those with the right education and training are qualified. Again, this is good news because you have either completed an educational program in electronics or are in the process of doing so—a fact that sets you apart from others in a positive way.

Regardless of whether you are pursuing or have completed a technical certificate, two-year degree, or four-year degree in electronics, you have chosen an excellent career field. As was mentioned earlier, electronics is a broad field that includes numerous areas of specialization. People with a solid education in electronics can be found working in positions that carry a variety of titles. The most widely used title in the field is *electronics technician,* but there are also numerous others, including aircraft electronics technician, avionics technician, automotive electronics technician, biomedical electronics technician, calibration/instrumentation technician, electronic communication technician, computer software technician, computer networking technician, computer hardware technician, electrical engineering technician, and many others. All of these positions are electronics technicians—they have just taken on more specifics relating to the area of specialization in question.

By enrolling in or completing an electronics program, you have made a good career choice. Now all that remains is for you to 1) complete your education if you are still in school, 2) secure the best possible position in your field, and 3) start climbing the career ladder.

Purpose of This Book

The purpose of this book is to help you with the two main challenges you will face upon completing an educational program in electronics. Your first challenge will be to secure the best possible job in your field. Your second challenge will be to succeed in your job and, over time, in your career. This book presents 12 strategies that will help you with both of these challenges. These 12 "success strategies" were developed on the basis of feedback from more than 200 professional and technical personnel interviewed by the author over a period of 3 years.

The people interviewed during the study were all either successful electronics technicians or management and supervisory personnel who hire electronics technicians. The consensus among these 200 study subjects was that the 12 specific success strategies presented in this book will help graduates of electronics programs secure good jobs as technicians and enjoy successful careers in the field.

The success strategies recommended by the study subjects interviewed by the author are as follows:

1. Master your school subjects.
2. Be a smart job seeker.
3. Build your career on a foundation of integrity.
4. Understand your employer's "big picture" and where you fit into it.
5. Apply self-discipline and time management.
6. Be a good team player, team builder, and team leader.
7. Be an effective communicator.
8. Be a critical thinker.
9. Learn to work well in a diverse environment.
10. Adopt a positive, "can-do" attitude toward your work.
11. Learn how to prevent and resolve conflict on the job.
12. Persevere when the job becomes difficult.

Each strategy in this list is explained in a chapter of its own. Each chapter is comprised of an explanation of why the strategy in question is important and specific actions you can take to implement the strategy. You can learn these success strategies in the same way you learned or are learning electronics skills in school. Study the material in this book in the same way you study your electronics materials in school and then begin applying what you are learning immediately.

Applying some of the success strategies in this book will be easier than applying others depending on your individual strengths and motivation. However, whether applying a given success strategy comes easily to you or requires extra work, with persistent effort, you can do everything recommended in this book and build yourself a successful career as an electronics technician.

ACKNOWLEDGMENTS

The author and Thomson Delmar Learning would like to acknowledge and thank the reviewers for their suggestions and comments. Thanks go to:

Surinder Jain
Sinclair Community College
Dayton, OH

Siben Dasgupta
Wentworth Institute
Boston, MA

Cree Stout
York Technical College
Rock Hill, SC

Mark Stewart
Sandersville Technical College
Tennille, GA

CHAPTER 1

MASTER YOUR SCHOOL SUBJECTS

LEARNING OBJECTIVES

Upon completion of this chapter, you should be able to do the following:

- *Explain why it is so important to master the courses in your school's electronics program.*

- *List the characteristics of the most successful students.*

- *Demonstrate how to apply study strategies to improve the effectiveness of your learning.*

- *Demonstrate how to apply selected strategies for enhancing your reading skills.*

- *Demonstrate how to apply selected strategies for enhancing your memorization skills.*

Your road to success as an electronics technician begins in school. Regardless of whether you study electronics at the high school, technical school, or college level, it is essential that you master your school subjects. What you learn in school will be the foundation upon which you will build your career in electronics.

Importance of Mastering School Subjects

Of the 12 success strategies presented in this book, mastering your school subjects is the most important. Unless you first do this, the other 11 strategies will not matter. A career is just like a house—it must be built on a solid foundation. For electronics technicians, that foundation is a solid grasp of the foundational principles of electronics taught in school.

What you learn in school will serve you well in two ways. First, it will be the basic body of knowledge and skills needed to practice your profession. This does not mean you will know everything you will ever need to know the day you begin work as an electronics technician. No matter how much or how well you learn in school, there will always be more to learn on the job. This fact accounts for the second way that what you learn in school will serve you well.

Electronics is a broad field consisting of numerous areas of specialization. No school program can teach everything there is to know in every one of these areas of specialization. This fact, coupled with the certainty that electronics is a field in which the technology and corresponding knowledge change rapidly and continually, means that you will have to be prepared to learn on the job over the course of your entire career. The foundational knowledge you develop in school is what prepares you to learn on the job for the rest of your career. Consequently, the importance of mastering your schoolwork in electronics cannot be overstated.

Your Electronics Studies

Because electronics is such a broad and varied field, the curriculums of electronics programs vary from school to school. Some programs take the more generic approach by teaching the fundamentals that apply to all electronics specializations. Others take the more specific approach by focusing on one or maybe two specific areas of specialization and going into more depth in these areas. Which approach your school takes is based on a number of factors, including the length of the program, the demands of local employers, and the overall purpose of the program in question.

Regardless of the length of your school's electronics program and regardless of the approach it takes to preparing students for careers in electronics, it is important that you learn well what your instructors teach. A program that takes the generic approach will teach you the fundamental principles of electronics you will need to know to get a job in your field and continue learning more specific principles on the job. A program that takes a more specific approach will also teach you the fundamental principles of electronics as well as how they are applied in a given area of specialization. Regardless of the approach your school takes to preparing you for a career in electronics, the most important thing you can do while in school is master your school subjects.

Characteristics of Successful Students

If you are not successful as an electronics student, you are not likely to be successful as an electronics technician. The attitude, learning strategies, and work ethic that make you a good student are the same as those that will make you a successful electronics technician. Consequently, as a student, you are not only learning electronics principles but also how to succeed.

Successful students can be different in a lot of ways—age, gender, race, height, weight—but in certain ways, they are all very much alike. Here are some of the characteristics typically displayed by successful students and how each characteristic will apply on the job:

- They attend all classes on a regular basis and are punctual. The workplace correlation of this characteristic is arriving at work on time or early every day.

- They go the *extra mile* in completing assignments rather than just doing enough to get by. The workplace correlation of this characteristic is going beyond just the minimum performance standards established by your employer and putting forth an extra effort.

- They pay attention in class and ask thoughtful questions when they do not understand. The workplace correlation of this characteristic is paying attention to your supervisor's instructions and asking clarifying questions when necessary.

- They try to complete assignments, finish projects, and solve problems themselves before asking the instructor for help. The workplace correlation of this characteristic is being a dependable employee who needs little or no supervision.

- They help fellow students who may not understand a given principle or who are having trouble with an assignment. The workplace correlation of this characteristic is teamwork—being a good team player.
- They give their best on all assignments, turning them in on time or early. The workplace correlation of this characteristic is the concept of *peak performance*—turning in the very best performance you can every day on the job.
- They accept the instructor's constructive criticism in a positive manner and use it to improve their work. The workplace correlation of this characteristic is the concept of *continual improvement*—accepting constructive criticism from your supervisor and using it to continually improve your performance on the job.
- They see the instructor in private before or after class if they disagree with a grade or any other aspect of their instruction. The workplace correlation of this characteristic is being a loyal employee who keeps disagreements with the supervisor private.

As you can see, each of these characteristics of successful students corresponds directly to a specific behavior on the job—a behavior that is associated with success. In other words, success on the job begins with success in the classroom. Experience has shown time and again that how you approach your studies in school is how you will most likely approach your work on the job. Re-read this list of the characteristics of successful students. Then, after reading each one, ask yourself: "Do I have this characteristic?" If you do not have a given characteristic, begin developing it right now.

Study Strategies That Will Enhance Learning

To master your school subjects and to continue learning over the course of your career, you need to know how to study (see Figure 1-1). This section contains a list of study strategies that can help enhance your learning during school and on the job.

- *Concentrate when studying.* When studying, block out all distractions, and focus on the material at hand. This means turning off the television, radio, or CD player and focusing exclusively on the material you are trying to learn.
- *Summarize what you are reading in your own words.* When reading a textbook, rather than highlighting what you think are the most important points, try this: stop reading after every major section, and summarize the main points in your own words. You may have to re-read to do this,

Checklist of:

EFFECTIVE STUDY STRATEGIES

- Concentrate.
- Summarize what you read.
- Establish a study schedule and stick to it.
- Make up potential test questions.
- Apply new principles immediately.
- Make use of supportive Web sites.
- Never skip over the difficult parts.
- Review your notes before each class.
- Study/review regularly—don't cram.
- Ask for help when it is needed.
- Test yourself periodically.
- Relate what you learn to the job.
- Do your best in all classes.
- Restudy material missed on tests.

FIGURE 1-1 These strategies will improve your learning.

but it represents the most effective way to ensure that you understand and will remember the material in question.

- *Establish a study schedule, and stick to it.* Studying your schoolwork should not be something you do when you have nothing else to do. Rather, studying should be a high-priority responsibility for which you block out specific times. Establish a schedule that blocks out sufficient time for studying your schoolwork, and protect that schedule—don't let anything that is not a bona fide emergency interfere with it.
- *Make up potential test questions.* After reading an assigned chapter in a textbook and summarizing the main points in your own words, complete one additional task: review your notes and make up potential test questions. Ask yourself: "If I were the instructor, what test questions

would I ask from this material?" Write those questions down, and make sure you can answer them.

- *Apply new principles as soon as you read or hear about them.* Any time the instructor presents a new principle—whether from one of your textbooks or during a lecture—apply it as soon as you can. You do not really understand a principle until you have applied it. For example, say you don't know how to ride a bicycle. You instructor presents an excellent lecture on how to ride a bike and even gives a demonstration in which he rides it around the room. Do you now know how to ride a bike? Of course not. The lecture and demonstration would be helpful, but until you actually climb on the bike and start trying to stay upright while pedaling, you haven't learned how to ride a bike.

- *Make full use of the Web sites that go with your textbooks.* Many of the more recent textbooks list Web sites that will help you expand your understanding of the principles and material presented in each chapter. Making full use of these Web sites will help you in two important ways: 1) it will add to the material covered in the chapter in question; and 2) it will get you accustomed to using Web sites for research and learning—something you will have to do on the job.

- *Never skip over the difficult parts of the work.* Some of what needs to be learned in electronics will come easily to you, and some will be more difficult to grasp. Students occasionally react to this fact by simply skipping the more difficult material. Never do this. The students who take the time and put forth the effort to master the more difficult material will be the most successful in school and on the job.

- *Review your notes before each class.* Before going to your next class, review what you learned in the previous class. This might mean reviewing the notes you made when reading a textbook, listening to a lecture, or watching an instructional DVD. What you learn today will be based on what you learned yesterday. Consequently, taking a few minutes to review what you learned in the previous class will ensure that you are prepared to learn effectively in the current class.

- *Study and review regularly rather than "cramming."* One of the most common practices of students is to put off studying for a test as long as possible and then "cram" the day or night before the test. This practice never works very well. The truth is, you can cram in only so much information in a short period without developing cognitive overload—what some students refer to as "brain lock." Regardless of what you call the concept, it is what happens when your mind has absorbed all it can in a short period of time. A more effective practice is to study on a regular schedule so that when a test is coming up, all you have to do is review your notes and you are ready.

- *Ask for help when having difficulty.* With all of your studies, make every effort to complete assignments on your own. Becoming an independent learner who is as self-sufficient as possible will serve you well when you have to learn on the job, and no instructor is available to help. However, when you have done all you can and still do not understand the material in question, ask for help. You might ask a fellow student who does appear to understand or your instructor. Never let the class leave you behind as it moves ahead to new concepts.

- *Test yourself periodically.* As you are learning new concepts, stop what you are doing periodically and conduct a self-test. Test yourself to see if you really understand a concept, remember a principle, or can apply a given skill. It's not enough to think "OK, I've got it—I understand." If you truly understand, you should be able to pass a self-test that proves you understand.

- *Try to relate what you learn in school to the job.* A practice that will enhance your learning and ensure that you are prepared for a job in your field is to relate what you learn in school to actual practices on the job. You will not always be able to do this yourself, but your instructor can. Whenever you learn new material, try to determine how that material would be used on the job or how it will be the foundation for additional material that will. If you have to ask your instructor for help in this regard, do so. Understanding the job-related correlations for your schoolwork will give that work more meaning.

- *Do your best—even in classes you dislike.* In any curriculum, you will like some classes more than others. The temptation is to put all of your effort into those you like while just doing enough to get by in those that you don't. This is a mistake. You need to put forth your best effort in all of your classes. The reason for this can be seen in the old coaching adage that says, "What you do in practice you will do in the game." In this adage, school is the practice element, and work is the game element. If you grow accustomed to just getting by in the classes you don't like in school, you are likely to do the same thing with the projects you don't like on the job. The most successful electronics technicians are those who do a good job on all of their work projects—not just those they enjoy. And, no matter how much you like your job in electronics, there will occasionally be individual work projects you don't like. Consequently, learning to give your best effort on all projects is important.

- *When you do poorly on a test, restudy the material missed.* It is going to happen—you are occasionally going to do poorly on a test. When this happens, go over the questions you missed, and restudy the material relating to those questions. Just because the test is over does not mean you don't need to know the material you missed. You are trying to master your coursework not just get through it.

Strategies That Will Enhance Reading

Mastering your school subjects will require a lot of reading, as will keeping yourself up-to-date on the job. Consequently, continually enhancing the effectiveness of your reading is a good strategy for becoming more successful in school as well as on the job. This section contains a list of strategies for enhancing the effectiveness of your reading.

Begin at the End

Before reading a chapter in a textbook, go to the end of that chapter and preview the questions, key terms, and other materials provided there. This will give you a good idea of what to look for as you read. One of the problems students face when reading school material is trying to understand what is important and what is less so. It is in the nature of textbooks to point out the most important material in a given chapter at the end of that chapter. The most important material might be pointed out in the form of a chapter summary, review questions, key terms, practice activities, or a combination of these. By previewing the end-of-chapter material before reading the chapter, you can determine what material deserves your special attention and what material is just "filler."

Develop a Chapter Outline

This step is easy to do if you are reading a textbook because chapters in textbooks are presented in an outline format. Before reading a chapter, flip through it and mark down all major headings on your outline and all subheadings as subheadings in the outline. Leave plenty of room between headings and subheadings for writing notes as you read. This will organize your reading and keep you from getting sidetracked on less important material.

Alternate Reading and Taking Notes

One of the most common approaches to reading for students is to highlight what they think is important as they read. Although this may be better than just reading the material without highlighting, it is not nearly as effective as taking notes. When you write out or even type reading notes, your memory imprints a mental picture of the words, thereby making it more likely you will remember what you have read. Just highlighting does nothing to form a mental image. Consequently, it is much better to make notes than to highlight.

However, you do not want to make notes while you are reading. Rather, try this approach. First, read a section in your book. Then put the book down and, using your outline as a guide, try to make notes for that section. At first, you will forget some of what you need to write down.

This is not a problem. Write down all of the important points you can remember, and then go back to the section in question and pick up any that you forgot. Repeat this process for each major section in the chapter. The more often you use this approach, the more material you will remember when putting the book aside and taking notes.

Develop Practice Test Questions

As you read, place a lightly penciled checkmark in the margin next to any information you think might make a good test question. In other words, if you had to test someone on the chapter in question, what material would you include in the test? After you have made your reading notes for a given section of a chapter, go back and add your potential test questions to your notes. Then ask yourself if—based on your notes—you can answer each question fully and accurately. If you can, mentally do so before moving on to the next section of reading. If you cannot, reread the material in question, and revise your notes accordingly.

Strategies That Will Enhance Your Memorization Skills

Mastering your school subjects and continuing to learn on the job will require you to remember a lot of material. In fact, memorization skills will serve you well throughout your career as an electronics technician. Much of the information you will need to know to complete your studies in school and do your job at work can be looked up in manuals as well as online but not all of it. There is always important material that is so universally applicable that you will need to memorize it and be able to recall it in a matter of seconds.

The more you can remember, the less you have to look up, and the less you have to look up, the faster you will be at doing your job. In a competitive workplace, time is always an issue. People who call on electronic technicians to diagnose, repair, assemble, install, or service electronic equipment seem to always want the job completed yesterday. Consequently, being able to do your work fast will be important. Being able to remember critical information will cut down on the amount of time your work takes. The following strategies will help enhance your ability to remember critical information.

Eliminate Distractions

Concentration is a key element in memorization. When studying your school assignments, eliminate all distractions and concentrate. This means that you should find a quiet place away from your friends, the television, radio, stereo, and any other distractions so that you can focus intently on what you are trying to learn and remember.

Engage Your Senses

Have you ever noticed that hearing a certain song will bring back a certain memory or that a certain smell will remind you of a favorite food dish? This is because the human memory responds to sensory stimuli. Consequently, the more of your senses you can bring to bear on material you need to memorize, the better. Reading the material will engage your sense of sight. Saying the material over several times out loud to yourself will engage the sense of hearing. Writing down material—taking notes as explained in the previous section—will engage the sense of sight again and reinforce your reading.

The senses of touch, feel, and taste can be more difficult to engage when studying electronics material, but if it is ever possible to engage these senses, do so. For example, if you are studying material about printed circuit boards and you can actually pick one up to touch it and even smell it, you are more likely to remember the material you are reading about circuit boards. As you study material that you want to memorize, try to find ways to engage as many of your senses as possible.

Make the Material Visual

Many people are visual learners in that they tend to remember what they see even better than what they read. Consequently, rather than just taking written notes when trying to remember reading material, try making quick sketches, drawings, diagrams, charts, and other visual aids. Doing so will implant a mental image in your mind that will be more easily recalled than just words—especially if you are a visual learner.

Form Memory Associations

A memory association is something that brings to mind something else. For example, smelling a turkey baking might bring to mind a favorite holiday meal you shared with family or friends. Hearing a particular song might bring to mind an enjoyable date you once had or a party you went to. Seeing a given letter might remind you of a whole word or phrase. Politicians use memory associations to remember the names of their constituents. For example, a politician might remember the name of a constituent named "Fred" because this person's hair is "red."

Forming a memory association involves connecting what you are trying to remember with something that you either already know or something that is easy for you to remember. For example, say you are trying to remember the names of the great lakes. If you can remember the word HOMES, it will help you—by association—remember the names of the lakes: Huron, Ontario, Michigan, Erie, and Superior. The association need do nothing more than give you a hint that will jog your memory. This is the key to developing memory associations. It is just a matter of finding little associations that will trigger the memory process.

Review Questions

1. Explain in your own words why it is so important to master the courses in your school's electronics program.
2. List and explain the characteristics shared by successful students.
3. List and explain the strategies presented in this chapter for improving the effectiveness of your studying.
4. List and explain the strategies presented in this chapter for improving your reading.
5. List and explain the strategies presented in this chapter for enhancing your memorization skills.

Discussion Questions

1. Refute or defend the following statement made by a student studying to be an electronics technician: "As long as I get by on Cs in school, I will do just fine when I get a job. I don't see why I should knock myself out trying to make better grades."
2. Refute or defend the following statement made by an instructor in an electronics class: "The reason I encourage you to do your best as students is that if you do your best in school, you are more likely to do your best on the job."

Application Assignments

1. Conduct a self-assessment. Make a list of the characteristics of successful students. When you have the list prepared, rate yourself from 1 to 10 with 10 being a perfect score on each of the characteristics. Then, pull any characteristic on which you score less than 8 out of

the list. This second list is your "Needs Improvement" list. Begin making the necessary improvements immediately.

2. Conduct another self-assessment. List all of the strategies for improving your study, reading, and memorization skills. Then rate the extent to which you apply each strategy using the 1 to 10 scale (with 10 being a perfect score). Pull any strategy on which you score less than 8 out of the list. This list is your "Learning Enhancement List." Begin applying the strategies on this list immediately.

CHAPTER 2

BE A SMART JOB SEEKER

LEARNING OBJECTIVES

Upon completion of this chapter, you should be able to do the following:

- *Compare your job-related needs with what potential employers will need from you.*
- *Develop your job-seeking portfolio (resume, list of references, cover letter, list of helpful sources).*
- *Prepare yourself for a job interview.*
- *Ensure an effective interview.*
- *Conclude an interview in a positive manner.*
- *Follow up after an interview.*

Before beginning preparations for the job search, step back and review your job-related needs as well as the needs of employers who might hire you. Smart job seekers make a point of pursuing not just any job, but jobs with a good "fit." A job with a good fit will meet your work-related needs while allowing you to help the employer meet its needs. So what are your needs with regard to a job, and what are the needs of employers who might hire you?

Your Job-Related Needs

Some of your needs will be very specific to you and you only (for example, wanting to work in a given location for family, health, or personal preference reasons). However, most of your job-related needs will fall into the following categories:

- Financial needs
- Personal needs
- Societal needs

Financial Needs

This should come as no surprise. Like most people, after completing school, you will need to make a living. You have chosen to do so by working as an electronics technician. This is an excellent choice because electronics is a high-wage career field, and the more you are able to earn, the better you will be able to provide for yourself and your family. Because money will be an important aspect of your long-term job satisfaction, it is a good idea to begin your job search by examining the prevailing wage and salary rates for electronics technicians in the geographic region where you plan to work.

How much are beginning electronics technicians in the region in question paid? You can find the answer to this question in any one of several ways. "Facts" sites on the Internet can be helpful sources of information. Want ads in local newspapers, friends or relatives who already work in the field, job listings on the Web sites of specific companies, and inquiries to the Human Resources offices of employers can also yield helpful information.

A good rule of thumb to remember when seeking a job is *never apply for a job without having a good idea of what the wage/salary should be.* You do not want to begin your career as an electronics technician by underselling yourself in job interviews. To ensure that you are paid what you are worth, you must first know what you are worth. As with any career field, what a beginner in electronics will earn can vary with geographic location and existing market conditions. Before applying for a job, you need to know what the prevailing wage rate is at the time and in the location in question.

Personal Needs

In a typical career, you will work as many as 40 years. Each year, you will work at least 50 weeks, and each week, you will work at least 40 hours. This means that over the course of a career, you will spend at least 80,000 hours working—and probably more. If you are going to spend this much time working, it will be important that you like your job. The most successful electronics technicians are satisfied with their jobs; they like their work.

It is important to understand this point because electronics is a diverse career field. You can do many different jobs under the auspices of being an electronics technician. Some people might be perfectly happy working as an electronics technician in the aerospace industry (avionics) but very unhappy working in the field of telecommunications and vice-versa. Some people might be happy making electronics products in a manufacturing setting, whereas others would be happier servicing those products. There are many different jobs under the broad heading of electronics that require essentially the same basic skills. In some of these jobs, you might be perfectly happy, and in others, you might not. The key, then, is to find an area of specialization within the broad field of electronics that is personally satisfying to you as an individual.

Societal Needs

In the long run, the people who are happiest in their careers are those who feel like what they do for a living is important—that it makes a contribution to the betterment of society. Much of your prestige in the eyes of society comes from what you do for a living. If people think that what you do is important, they will think that you are important.

When you are just beginning your career as an electronics technician, money and personal satisfaction will probably be your highest priorities. However, over time as you advance up the career ladder—especially if you are earning a respectable income in a job you find satisfying—knowing that what you do for a living matters will become increasingly important to you.

This is just one more reason why choosing electronics as your career field was a wise decision. We live in an electronic world where society depends on electronics technicians to keep critical technologies working. Without you and your colleagues in this important field, automobiles will not drive, airplanes will not fly, medical equipment will not operate, telephones will not send or receive, televisions will not work, and on and on. The list of contributions made to society by electronics technicians is a long one.

As you begin preparing for the job search, remember that you will want to find a job that is a good fit. This means you will want to find a job that pays what you are worth, gives you personal satisfaction, and provides you an opportunity to make a contribution. In the field of electronics, accomplishing this should not be difficult after you have completed your education.

Your Employer's Needs

Much of the rest of this book is geared toward helping you make yourself more successful by helping you make your employer more successful. In this section, the needs of employers that might hire you are explained in "big-picture" terms. It is important for you to understand these needs because the most successful electronics technicians are those who do the best job of meeting their employer's needs.

The most satisfying situation is one in which the job allows you to meet your work-related needs—financial, personal, and societal—while simultaneously helping meet your employer's needs. This is the concept of *mutual benefit,* and it represents the ideal employment situation.

Regardless of whether your employer is a private sector company, a public sector organization, a nonprofit agency, or a military unit, its business will involve providing a product, a service, or both. Because most employers now operate in a globally competitive environment, the more effectively and efficiently they produce their products or provide their services, the more successful they will be.

Consequently, your employer needs two important things from you: 1) consistent peak performance and 2) continual improvement. Consistent peak performance means that you do your absolute best work every day, week after week, month after month, and year after year. Continual improvement means that you improve your performance on the job all the time forever. In other words, you never stop getting better and better at what you do.

By turning in peak performance on a consistent basis and by improving your performance continually, you help your employer operate more efficiently and effectively, which, in turn, makes your employer more competitive. An organization is just like a baseball team in that the better each member of the team plays, the better the organization performs overall. Once again, this is the concept of mutual benefit—the ideal employment situation.

Developing Your Job-Seeking Portfolio

Just as you enrolled in an electronics program to prepare yourself for a career as an electronics technician, you should develop a job-seeking portfolio to prepare yourself for the job search. The better your preparation in developing a job-seeking portfolio, the better the job search will go. Your portfolio should contain all of the materials you will need to conduct an effective and ultimately successful job search.

A comprehensive job-seeking portfolio for an electronics technician contains the following items:

- Resume
- Sample cover letter
- List of references
- List of helpful resources

Developing Your Resume

The resume will be one of your most important job-seeking tools. Consequently, it must be properly developed; which is to say that it must be just the right length, be attractive, and be informative. Before beginning to actually develop your resume, you need to understand what one is and how it will be used—by you and by potential employers. Think of your resume as a marketing brochure and the product being marketed is YOU. From your perspective, a well-developed resume will do the following:

- Explain your career goal and how it "fits" with the goals of organizations that hire electronics technicians.
- Summarize your qualifications in a way that shows employers you have the potential to help them.
- Help get your foot in the door for an interview.
- Represent you with the employer after the interview but before a hiring decision has been made.

From the employer's perspective, your resume will 1) provide a tool for deciding whether or not to grant you an interview, 2) provide a glimpse of your personal characteristics (for example, neatness, accuracy, attention to detail, preparation, and so on), and 3) provide a "picture" of your qualifications relating to the job in question.

When an employer has more applicants than openings, your resume becomes especially important because it will be used to "weed out" applicants who appear less qualified or who appear equally qualified but lack certain desirable personal characteristics (as mentioned in the previous paragraph).

Characteristics of a Well-Written Resume

Assume you want to buy a car but are unsure what kind and model you want. You have a good idea of what general features you want and what you can afford. However, many different cars would satisfy these needs. Rather than waste time walking around car lots and being pestered by salespeople, you decide to begin your search online. Your plan is to go online and read the information provided about various cars in your price range. Based on this information, you plan to eliminate all cars that do not fit your profile, thereby narrowing the list of possibilities down to those you want to actually see in person and test drive.

The better the job done by automobile manufacturers in preparing and presenting their online information package, the more likely it is that you will put their car on your "test-drive" list. For example, there might be a car that falls in your price range and has some of the general characteristics you are looking for. However, because the automobile manufacturer has done a poor job of putting together its online information, there are questions about this car that you cannot answer. Additionally, some of the information provided is inaccurate and incomplete. Consequently, you eliminate this car from your list of possibilities.

In this analogy, the online information package presented by each automobile manufacturer is its "resume." The test drives are the interviews. The job search is a lot like this car-buying analogy. The better the job you do in preparing your resume, the better your chance of making an employer's "short list" for an interview. To improve your chances of making an employer's short list, make sure your resume has the characteristics described in the following sections.

Future-oriented

Your resume should show potential employers where you would like to go in your career, not just where you are now or where you have been. This is accomplished by developing a well-crafted career goal. A career goal is a simple statement of where you would like to go in your career. Here is an example of a career goal for an electronics technician who wants to work in a manufacturing setting:

To begin work as an electronics technician in a manufacturing setting and rise in my field over time to the level of division manager.

Notice that this goal shows potential employers that you want to start out as an electronics technician, that you want to work in a manufacturing setting, and that over time you plan to work your way up the career ladder and become the manager of the electronics division of a company. This is a simple but powerful statement. It shows that you are willing to start at the bottom and work to gain experience, that you know what specialty within the broad field of electronics you are interested in, and that you have the ambition to work your way to the top in your field. Employers reading your resume would view all of this information positively.

Brief and easily read

Employers are always busy. Consequently, they are unlikely to read more than one or two pages of resume material. For this reason, it is important to limit the length of your resume. At this point in your career, one page is preferable and two pages acceptable, but more than two is too many. This is why developing an effective resume can be such a challenge. It must say a lot in relatively few words.

Balanced between general and specific

A well-written resume will summarize your qualifications in terms that are sufficiently general in nature to encompass a range of jobs in the broad field of electronics but are specific enough to show you can actually do the job. This is a real balancing act that can require several revisions to your resume before getting it right.

Here is a situation that sometimes happens when looking for a job. You respond to an advertised job by sending a resume to the employer in question. Someone else gets the job that was advertised, but after reviewing your resume, the employer finds that you might qualify for another job that is about to be advertised. Rather than advertise the new job opening, the employer calls you in for an interview. In this case, because you took the time to develop and present a good resume, you got a chance to interview. This type of situation is why it is important to balance general and specific information in your resume.

In this example, if your resume had been too specific, the employer would not have known that you could do the other job. Correspondingly, had it been too general in nature, it probably never would have been considered in the first place. The key is to include information in your resume that shows you can do a range of different jobs under the broad heading of electronics, while also giving enough detail to show that you are actually qualified.

Written in the terminology of the field

Electronics is like any career field in that it has its own language and terminology. Resumes are collected by human resource management personnel, but they are screened in detail by experienced electronics technicians and their supervisors. Most companies that advertise openings for electronics technicians will convene a selection committee consisting of human resource management personnel, electronics technicians, and the individual who will supervisor the position in question. This means that most people who will review your resume will be electronics professionals who know the language and terminology of the field.

One of the best ways to make a favorable impression on the electronics professionals on the selection committee is to use the language of the field in writing your resume. Before writing your resume, do some research. Visit the Web sites of potential employers and make note of the terms they use when advertising openings for electronics technicians. If there is a job announcement or advertisement for a position you would like to have, make note of the terminology used in writing it, and "parrot" that terminology in your resume. This will ensure that you are speaking the language of the people who will review your resume.

Types of Resumes

There are several types of resumes, each with its own advantages and appropriate applications. Of these, the most widely used—and the best for electronics technicians—are as follows:

- Categorized capabilities
- Chronological
- Combined

Each of these types of resumes has a more appropriate application depending on the relative qualifications and experience of the individual in question—you. It is good to understand the relative merits of each type before deciding which to use.

Categorized capabilities

This type of resume works well for people who do not yet have any experience working as an electronics technician. It consists of the following elements: 1) career goal, 2) education summary, and 3) a series of capability summaries by category. It is based on the resume writing principle that says: "If you cannot show employers what you have done (experience), show them what you can do (capabilities)." Your capabilities are the knowledge and skills relating to electronics that you developed in school. Figure 2-1 is an example of a categorized capabilities resume.

Chronological

This type of resume works well for experienced electronics technicians. Although few students of electronics are going to be experienced electronics technicians, some are. Some students work in the field and pursue their education at night. Others work part-time in the field while completing their studies. If either of these situations apply, the chronological resume is appropriate. It consists of the following

John Jones
15 North Oak Street
Ocala, Florida 32398
904-678-5151 jjones@qrt.net

Career Goal
To begin work as an electronics technician in a troubleshooting, service, and repair position and rise over time to the level of electronics service manager.

Electronics Capabilities
In completing a technical certificate program in electronics technology, I developed entry-level knowledge and skills in the following areas:

- *Fundamentals.* Direct current, alternating current, impedance, and power supplies.

- *Wired electronics.* Semiconductor diodes, transistors and integrated circuits, signal amplifiers, and signal oscillators.

- *Wireless electronics.* Radio-frequency transmitters, radio-frequency receivers, telecommunications, and antennas.

Education
Technical Certificate in Electronics Technology, Macome Center for Applied Technology, May 2007.

High School Diploma, Macome High School, June 2006.

FIGURE 2-1 Categorized capabilities resume.

elements: 1) career goal, 2) education summary, and 3) summaries of work experience beginning with the most recent and working backwards to the least recent. Figure 2-2 is an example of a chronological resume.

Combined

This type of resume works well for people who have some work experience—in the field of electronics or in other fields—but not a lot. It contains the following elements: 1) career goal, 2) education summary, 3) a series of capabilities summaries by category, and 4) summary statements of work experience. If you have some work experience in electronics or even relating to electronics—regardless of whether it is part-time or full-time—list it first, even if this is out of chronological order. Figure 2-3 is an example of a combined resume.

Gabriela Rodriguez
19 South 15th Street
Miami, Florida 45678
321-898-965   grodriguez@mnp.net

Career Goal
To continue my career in electronics in a manufacturing setting that will allow me to apply my knowledge and skills and advance at a rate that is commensurate with my performance on the job.

Experience
A total of five years of experience working as an electronics technician in the military and the private sector:

- *United States Army.* January 2002–December 2005. Served as an electronics communication specialist.
- *Miami-Dade College.* January 2006–December 2007. Worked part-time (20 hours per week) as a Computer Service Technician and Laboratory Assistant.

Education
Associate of Applied Science Degree in Electronics Technology, Miami-Dade College, December 2007

High School Diploma, Miami Central High School, June 2001

FIGURE 2-2 Chronological resume.

Developing Cover letters

When responding to job advertisements, never send just a resume. Instead, attach your resume to a cover letter. The letter and resume may be sent by e-mail or the old-fashioned way. But either way, always send a cover letter. The cover letter serves three important purposes:

- Lets the employer know what specific job you are interested in
- Points out specific details in your resume that show you are qualified for the job in question
- Allows you to tailor your application to the specific job in question without having to completely rewrite your resume every time you apply for a different job

Nheiu Vo Din
Box 15, Rural Route 21
Pennville, Nebraska 87690
567-897-9534   vodin@ade.net

Career Goal
To begin work as an electronics technician calibrating, servicing, and repairing medical equipment and to rise to a position of CEO of my own electronics service company.

Electronics Capabilities
In completing a Bachelors Degree in Electronics Technology, I developed knowledge and skills in the following areas:

- Testing, calibrating, operating, and repairing electrical, electronic, and electro-mechanical devices, apparatus, and equipment.

- Performing routine maintenance and troubleshooting on equipment and devices, determining what types of repairs are necessary, and making those repairs.

- Conducting tests and inspections on equipment and devices to evaluate the quality of their performance.

Experience
A total of four years of part-time experience working while pursuing my college degree:

- *Sales Clerk.* Radio Shack, January 2004–December 2005.

- *Computer Lab Assistant.* University of West Florida, January 2006–December 2007.

Education
Bachelor of Applied Science Degree in Electronics Technology, University of West Florida, December 2007.

High School Diploma, Escambia High School, June 2003.

FIGURE 2-3 Combined resume.

The overall purpose of the cover letter is to make sure your resume is read rather than just tossed in a trash can. It does this by letting the employer know what job you are interested in, pointing out specific qualifications you have relating to the job in question, and highlighting

Date

Dear _____:

I am applying for the position you advertised for an Electronics Technician. I have a Technical Certificate in Electronics Technology from Escambia Applied Technology Center and am ready to begin work immediately in the type of position in question. Please allow me to highlight some of my specialized training and education as it relates to your position:

- You are looking for someone who has skills in the areas of electronic troubleshooting and equipment failure diagnosis. My final course in school covered troubleshooting and diagnosis in depth.

- You are looking for someone who can work well with customers when making calls in their facilities. I have strong people skills and am able to get along well with strangers in any setting. Also, my training at Escambia Applied Technology Center stressed the need for good customer-service skills.

- You are looking for someone who can work on his own with little or no supervision. I am able to work well in an independent setting where I have to make decisions, set schedules, and supervise myself. Evidence of this can be seen in my transcript that shows I completed several of my courses in school by distance learning (online). This required self-discipline and self-supervision.

I will appreciate an opportunity to interview for this position so that I can offer additional evidence of my ability to do an excellent job for your company.

Respectfully,

John Smith
302 N. Oak Avenue
Niceville, Florida 32578
850-975-8856

FIGURE 2-4 A well-written cover letter can help you land an interview.

those parts of your resume that relate most closely to the job in question. The cover letter should contain the following elements (see Figure 2-4):

- Reference to the specific job you are applying for
- A brief summary of your qualifications as they match selected requirements from the job advertisement

- Your contact information (telephone number, address, e-mail address, cell phone number, and so on)

The cover letter, if properly written, will call the employer's attention to your most pertinent qualifications relating to the job in question. It will also allow you to tailor your application to the specific job in question without having to rewrite your resume to match the advertised qualifications for a given job. This tailoring aspect of the cover letter is important because you have already gone to great effort to achieve an appropriate balance between generality and specificity.

Of course, you could rewrite your resume and tailor it to every specific job you apply for, but in addition to being extremely time consuming, this approach might actually eliminate you from consideration for a job. As was explained earlier in this chapter, sometimes there are jobs available in organizations that have not yet been advertised or that were advertised but were not filled. By making your resume too specific to a particular job, you might rule out being considered for one of these positions. Your resume should achieve an appropriate balance between generality and specificity, but your cover letter should be tailored to a specific job that has been advertised.

Developing Your List of References

After you have your resume developed and a good sample letter of introduction on file, it is time to develop a comprehensive list of references. Very few employers will hire you without first checking your references. References are people who can attest to your character, work-related qualifications, and schoolwork. When choosing references, think of people in the following categories who will speak well of you: personal, work-related, and school.

Personal References

Personal references are people who can attest to your character. Employers will want to know if you are honest, dependable, and trustworthy. People who have interacted with you on a personal basis for an extended period of time—the longer the better—make the best personal references. Remember this about choosing personal references: you want references who will command the respect of potential employers. Consequently, the best personal references are people who 1) have known you long enough to have a credible opinion of your personal character (five or more years is best) and 2) hold sufficiently prominent positions in the community to command the respect of potential employers (coaches, clergy, elected officials, teachers who know you on a personal basis, business leaders, military officials, and so on).

Work-Related References

If you have a job of any kind right now or if you have ever have had a job, employers will want to talk with a supervisor and/or fellow employee who can attest to your work ethic and work habits. Employers will want to know if you have a positive attitude toward work in general as well as if you are punctual, cooperative, and a good team player. They will want to know if you are able to function well in a diverse environment and under the pressure of deadlines.

These traits are universally applicable—they are important in every job, not just that of the electronics technician. Consequently, the type of job or jobs you have had—full-time or part-time—is not as important when choosing references as identifying supervisors who will speak well of your work ethic and work habits.

School References

Employers will want to know if you are a capable learner who completes assignments on time, if you take the initiative to go beyond just minimum requirements, and if you give your best effort on all projects. They will also want to know if you are able to work well with your fellow students. Instructors, counselors, and administrators from your school are in a good position to attest to these school-related characteristics, and they typically have a high level of credibility with employers.

Securing the Permission and Cooperation of References

Never make the mistake of putting people on your list of references without first obtaining their permission. After you have their permission, go one step further and obtain their cooperation. In other words, never assume that you know what a given reference will say about you. Make sure. Visit each person on your list, secure the necessary permissions, and then give each reference a typed list of the characteristics you would like stressed when he or she talks with potential employers. Ask all of your references for their cooperation in using your list of characteristics when called by employers.

This second step—securing cooperation—is critical. When seeking a job in your field, leave as little to chance as possible. Never assume you know what references will say to employers or that they will know what to say. Also you never know how employers will interpret what references mean by what they say. Sometimes a comment meant as a joke by one of your references will be taken seriously by employers. Consequently, it is important to give your references a "script" (the list of characteristics you want them to emphasize) and ask them to stick as close as possible to the script. The reference checking process should be like a carefully choreographed dance, and *you* are the choreographer. Figure 2-5 is an example of a list of characteristics you might give to references.

Employment Characteristics

For

Joe Smith
413 Elm Lane
Fort Walton Beach, Florida 32547
850-897-2435   smithj@xyz.com

- Punctual

- Dependable

- Honest

- Team player

- Leader

- Positive attitude

- Cooperative

- Flexible

- Perseverance

- Learner

- People skills

- Diversity-friendly

- Vital/energetic

- Supervisable

FIGURE 2-5 The type of list you might give to your references.

You do not want every reference to attest to every characteristic on your list. Rather, talk with each reference and come to an agreement concerning the characteristics he or she will focus on. This approach will ensure that all of the characteristics on the list are spoken to by at least one of your references, but that all references are not saying the same thing.

Arranging Secondary References

After you have secured the permission and cooperation of your primary references—those you will provide to employers when they ask—go one step further: arrange secondary references. Secondary references are people your primary references are asked to provide to employers during the reference checking process. It works like this. You provide a list of references to an employer. This is your list of primary references. The employer calls the references you provide and, during these calls, asks each reference to provide the name of someone else who knows you.

Not all employers do this, but those that do are being cautious. For example, if you are applying for a job with the government or a Department of Defense contractor where security is an issue, employers often use this practice. Some employers expect the references you provide to say good things about you—otherwise, why would you provide them? Consequently, to make sure they are getting a less biased point of view, they ask your primary references to provide other references as a way to gain a second, less-scripted opinion.

Because some employers use this practice, wise job seekers simply add identifying secondary references to their "to do" list when preparing for the job search. All you have to do is give each primary reference the names and telephone numbers of one or two other references. Then tell each primary reference they might be asked by employers to provide secondary references. Tell each, if asked by employers for secondary references, to give only the names you have provided.

Once again, do not assume either the permission or cooperation of secondary references. Treat secondary references just like primary references. Ask for their permission, secure their cooperation, and provide them with a list of characteristics like the one in Figure 2-5. Come to an agreement concerning which characteristics each secondary reference will emphasize.

Developing Your List of Resources

An important component of your job search portfolio is a list of resources that will be helpful in locating a job that is right for you. These resources will include the Internet, newspapers, employment agencies, and people you know who might be able to open doors.

The Internet as a Resource

The Internet can be one of your best resources when looking for a job. There are now hundreds of employment Web sites you can use to search for jobs and to broadcast your resume to potential employers. These are the "big three" employment Web sites:

- Careerbuilder.com
- Monster.com
- Hotjobs.com

Although these are the largest and best known employment Web sites, there are many others. Using Internet Web sites involves 1) going to the site; 2) following the directions for registering at each site; and 3) posting you resume to the Web site's job board. The instructions for accomplishing these tasks will vary from site to site. Consequently, it is a good idea to begin visiting employment Web sites while still in school to become familiar with their individual requirements.

Some employment Web sites ask you to electronically cut and past your resume to load it on the site. Others ask you to complete an extensive questionnaire that is then used to match your qualifications with job requirements. In either case—cut and paste or questionnaire—what you put on the site will determine what you get out of it. When you input your qualifications at an employment Web site, it is important to 1) take the time to be comprehensive—don't leave out anything that might help you and 2) be smart about how you say what you say. This second strategy can be critical in some cases because employers will use the information you provide as much to eliminate you as a candidate as to accept you for further consideration.

When entering information onto an employment Web site, try to put yourself in the shoes of employers who will read the information you provide. It is important to strike an appropriate balance between generality and specificity as you did when developing your resume. Again, an appropriate balance is one in which your information is general enough to connect you to as many potential jobs as possible but specific enough to show that you are qualified. Following is a list of employment Web sites that may be helpful in your job search:

- www.careerbuilder.com (all jobs)
- www.hotjobs.com (all jobs)
- www.monster.com (all jobs)
- www.flipdog.com (all jobs)
- www.americasjobbank.com (all jobs)
- www.careersite.com (all jobs)
- www.jobbankusa.com (all jobs)
- www.aerojobs.com (aerospace and aviation jobs)
- www.tvandradiojobs.com (television and radio jobs)
- www.techemployment.com (computer-related job)

Applying Directly to Employers

Another benefit of the Internet is that it provides direct access to employers that might have just the right job for you. These days, even the smallest companies maintain a Web site. Consequently, if you know the name of a company, finding its Web site is a simple matter of conducting an online search based on the company's name. If you do not have a specific company in mind and need to identify several, simply conduct a search based on industry type and location. For example, you might key in manufacturing companies in Orlando or avionics companies in Chicago or medical device companies in Omaha.

After you have identified companies in your preferred industry sector and location, begin visiting their respective Web sites. Most employers use their Web sites to advertise job openings and recruit applicants for those openings. Consequently, with just a few clicks, you can determine what openings an employer has and how to apply.

It is a good idea to begin visiting the Web sites of specific employers while still in school so that you can graduate already knowing how to work the system and what to expect from it. By visiting applicable Web sites before beginning your job search, you can also learn the language of the job search and key terms to include in your resume and letter of introduction.

Newspapers

Even in the age of computers, the newspaper can be a helpful resource for finding a good job in your field. The best day of the week to check the want ads in a newspaper is Sunday because more people read the Sunday newspaper than any other day. Knowing this, employers ensure maximum exposure for their want ads by running them in the Sunday edition.

If you plan to relocate to another community after graduation, newspapers for that community can be obtained in several ways. One way is to purchase the Sunday edition of the newspaper from the community in question from a newsstand or from the newspaper racks found in front of most larger grocery stores. Another source is the local public library or the library of your school. You might ask a friend or relative who lives in the community in question to send you the Sunday want ads, purchase a short-term subscription to the newspaper in question, or download the want ads if the newspaper in question is available online.

Remember when reading want ads, you should make a point to review back issues of the newspaper because technical jobs such as electronics technician often go unfilled even though the employer's ad is no longer running. Some employers get frustrated when their newspaper ads fail to produce results and simply drop the ad without filling the job. Others decide from the outset to advertise for only a predetermined period of time—say two weeks. If the ad fails to produce results during the prescribed period, they drop the ad and try another approach. Consequently, it is not uncommon for an outdated ad to still be a valid lead.

Employment Agencies

Usually public and private employment offices are available in all but the smallest of towns. In fact, your school, college, or university may have its own placement office. If there is a state-funded employment agency in your community, add it to your list of job search resources, and visit the nearest office while still in school. By visiting the office now, you can learn how it operates before graduation. Then, if you need help from this agency, you will already know how it works.

Private employment agencies are another matter. Before adding a private employment agency to your job search resource list, it is a good idea to ask a few questions of the personnel there:

- Is there a fee for the services provided?
- If there is a fee, who pays it—you or the employer? (with your education and training, you do not need to pay a fee to find a job)
- Does the agency connect clients (you) with permanent jobs or just temporary ones?

The only private employment agencies that should be added to your job search resources list are those that require employers to pay any and all fees and that refer you to permanent jobs.

If your school, college, or university has a job placement center, visit the center frequently while still in school. Learn what services are available and how to use those services. Employers often work closely with school-based job placement centers listing job openings with them and even conducting on-site interviews.

Job Fairs

Recruiting companies, organizations such as local Workforce Development Boards, and individual employers sometimes combine resources to conduct job fairs. A job fair is an event in which employers or their representatives set up booths in a convenient location, such as a conference center, school, college, university, or auditorium, and advertise to attract people who are looking for jobs. Each booth will have information available about job openings with the employer in question and how to apply. Some employers will actually conduct preliminary interviews on the spot. If you attend job fairs while still in school or after graduation—and you should—apply the following strategies:

- Dress up as if you are going to an interview because you might be.
- Take along several copies of your resume in a stiff folder that will protect them against wrinkles and smudges.
- Talk to employers in the smaller booths too. Some employers invest in building an attractive booth with lots of "bells and whistles." Others simply set up a table and drape a sign over it. Remember this—a better booth does not necessarily mean a better employer. Stop at all booths that advertise jobs in electronics—the small and the large.
- Collect business cards from every employer you talk to, even if there is no job available with a given employer at the moment. You might want to contact an employer again in the future.

- Collect flyers and brochures from employers that interest you. The information provided in flyers and brochures will be helpful in preparing you for interviews with the employers in question.

Some Web sites can help you determine when and where job fairs are scheduled. In fact, some job fairs are actually held online now. While still in school, you may want to visit the following Web sites to familiarize yourself with job fairs as a resource:

- www.careerfairs.com
- www.cfg-inc.com
- www.psijobfair.com

Preparing for an Interview

All of your preparation up to this point—resume, cover letter, list of references, and list of helpful resources—was done for the purpose of securing an interview. In order to get a job, you must first get an interview. The interview is the most important step in the job-seeking process—it is the "make-it-or-break-it" step. Consequently, you must be well prepared for interviews. The following preparation strategies will help enhance your chances of winning the job during an interview:

- Understand what employers are looking for during an interview.
- Know how to dress for an interview.
- Know how to send the right nonverbal signals.

What Employers Are Looking for in an Interview

Before inviting you to an interview, an employer has already formed an opinion of your technical qualifications by reviewing your resume and cover letter. The resume and cover letter have already impressed the right people sufficiently to secure an interview for you. Consequently, you should expect that during the interview, the employer is going to look for more than just your electronics-related qualifications. In addition to the hard skills relating to electronics, interviewers will also be looking for evidence of the following *soft skills*:

Fit. Just as you will want to select the job that is best suited to your individual needs, employers want to hire the technicians who are best suited to their needs. This is the concept of *fit,* and it applies to both you and employers. Every organization has a culture—a way of doing things, an attitude toward work, ways in which people interact and relate to each other, and unspoken expectations. Employers work hard to establish and maintain the desired organizational culture. Consequently, they want to hire those technicians who will fit into their culture.

Willingness. During interviews, you might be questioned about your willingness to do what technicians often refer to as "grunt work." Grunt work consists of menial tasks that must be done—making coffee for the team first thing in the morning or running errands—but that do not require the skills of an electronics technician. Such tasks are often assigned to the newest technician on the team. In my first job as a technician, one of my grunt-work tasks was to stop by a local bakery every Friday morning and pick up donuts for the team. We did this on Fridays only.

To compensate me for my trouble, I did not have to put any money into the "donut fund" during the week. The other team members paid, and I picked them up. Employers who ask about your willingness to do grunt work may just be testing you. They want to know if you are willing to start at the bottom and pay your dues as part of the process of earning your place on the team. This is important. Almost every team of any kind expects its new members to undergo certain rites of passage before becoming fully accepted as part of the team. This is just human nature.

Experienced team members think, "I had to go through it—let's see if you are willing to do what I had to do." You might never be asked to actually perform any menial tasks for the team, but it is important to show that you are willing to do so. If asked to do grunt work, just remember, you will not be the new person on the team for long. As soon as someone else is hired, he or she will take over the menial tasks.

Teamwork. As an electronics technician, you will work as part of a team. Your supervisor is the coach, and your fellow employees are your teammates. Consequently, employers will want to know that you can 1) be a good team player, 2) work well with others, 3) put the team's needs ahead of any personal agendas you might have, and 4) work well in a diverse environment. Remember that your answers to questions in interviews should indicate that you can work well in a team and that you will do your part to make the team better.

There is a story from college basketball that illustrates the importance of being a team player—even if you are the star of the team. There was once a famous college basketball player who was intelligent, informed, and fiercely independent. He was also the "franchise" player on his college's basketball team—the player whose talent could bring the team a national championship.

This particular team was coached by a soft-spoken man who was patient with the individual differences of his players and encouraged them to think for themselves—that is until their behavior clashed with his team rules. One of those rules was no players with beards or long hair on his team. This was in the early 1970s when almost all college students wore long hair and many wore beards.

After a long break between terms, the coach's franchise player—the player whose talent could mean a national championship for the team—showed up for practice with the beginnings of a beard he was growing. His teammates, knowing the coach tolerated no breach of team rules, waited in suspense to see what the coach would do. After all, this was the best player in college basketball at the time.

Without making a big fuss, the coach simply waited for a break in practice and then pulled the bearded player aside and reminded him about the team's rules on facial hair. The player countered that whether or not he wore a beard was a matter of personal choice and that he had a right to wear one if he wanted to. The coach acknowledged that the player was correct and that he did, in fact, have every right to wear a beard if he chose to. Then he wished the player good luck and told him the team would surely miss him. In other words, the player could keep his beard but at a cost of forfeiting his place on the team.

The coach's point was clear. Every player on the team had a right to make his own personal decisions, but decisions have consequences. If exercising your individual rights or pursuing your individual agenda puts you at odds with what is best for the team, you can make your own choices, but there will be consequences. You cannot be part of a team and simply go your own way when you don't like the direction the team is going. You may certainly recommend a different direction and attempt to persuade the coach (your supervisor) of the benefits of your recommendation, but you cannot just go one way while the team goes another and still be a member of the team.

Supervisability

Can you take direction and instructions well? Will you? Do you respond positively to constructive criticism? Are you willing to follow your supervisor's instructions without explanation when circumstances dictate the need—such as in a crisis—as long as what you are told to do is legal and ethical? Technicians who can answer "yes" to these questions are easier to supervise than those who cannot.

Since your potential new supervisor will probably be on the selection committee that interviews you, supervisability will be an important factor. No supervisor wants to hire a problem or a maverick who will go his own way regardless of what is best for the team. Consequently, during interviews it will be important to give answers that reflect a willingness to be supervised.

Critical Thinking

On the one hand, employers will want you to be a team player who is easily supervised or, better yet, who requires little or no supervision. On the other hand, they do not want to hire sycophants or robots. The best technicians are not just easily supervised team players, they are also critical thinkers. This means they approach their jobs with intelligence and with a mind to always looking for better ways to get the job done. Peak performance and continual improvement are always foremost in the minds of successful electronics technicians.

If electronics technicians who are critical thinkers see ways to do some aspect of the job better—meaning more efficiently, more effectively, or in any way that will improve quality, cost, or service—they speak up. This is done in a respectful manner and might even be done in a private conference with the supervisor, but it is always done. Critical thinkers never just go along with the way it's always been done when they see a better way.

Dressing Appropriately for an Interview

An interview is like a first date—you really want to make a good impression. Consequently, how you dress for an interview is important. Remember, the first impression you make will be based on how you look—and good first impressions are critical because they tend to be lasting. The first impression you make is the one most people will remember. Dressing properly for interviews will help you make a good first impression and sends several nonverbal messages, including the following:

- You respect the interviewers and the interviewing process enough to take the time to dress well.
- You care enough to want to make a good impression.
- The interview is important to you.

Dressing for Interviews: Men

The first rule in dressing for an interview is to dress up in a conservative way—nothing flashy. Even if the organization in question has a business casual dress code, men should wear either a suit or a coat and tie to the interview. Dark sports coats or suits are best—but do not wear black. Dark blue, gray, or brown solids are recommended for interviews. Your shirt should be long-sleeved and either white or another light color (blue, beige, ecru, and so on). Your tie should complement the color of the suit or sports coat but should not be loud. Always, when dressing for an interview, dress in an understated and conservative manner.

Shoes should be dark—black or brown—leather dress shoes. Make sure your shoes are clean and shined. Socks should be dark and match the suit or sports coat color. The only jewelry you should wear to an interview, if any, is a watch and a ring, but no more than one ring per hand and nothing flashy or gaudy—remember to be understated and conservative. If you typically wear bracelets, earrings, nose rings, neck chains, or any other kind of jewelry other than a watch or ring, leave them at home during interviews.

Dressing for Interviews: Women

For women, the first rule in dressing for an interview is the same as for men: dress up, but be conservative. Conservatively cut skirts, dresses, jackets, and suits in dark colors other than black (blue, gray, brown) are all appropriate for interviews. Hemlines should fall right at the knee or no more than two inches above the knee.

Blouses should be long-sleeved and of a solid color that complements your skirt, dress, suit, or jacket (white, beige, light blue, ecru, and various subtle pastels). Shoes should be dark in color and leather. The key is to select shoes that complement your outfit and are as dark or are darker than your skirt, dress, jacket, or suit. Hose should be neutral skin tones. Belts, if worn, should match your shoes.

The best advice concerning jewelry is to wear very little during interviews and to be conservative with what you do wear. When interviewing, wear no more than one ring per hand and only rings that are understated— nothing gaudy. If you wear earrings, make sure they are small and conservative. All other jewelry should be left at home when interviewing. Makeup worn to an interview, like clothing, should be conservative and understated. Take an understated approach with both eye makeup and lipstick; for interviews, less is better.

Sending the Right Nonverbal Signals

Much of human communication is nonverbal. We send hundreds of nonverbal signals with our handshakes, posture, facial expressions, eye contact, rate and tone of speech, gestures, and proximity. Following are strategies for sending the right nonverbal signals during an interview.

- *Handshake.* Much can be communicated by a handshake—good and bad. You want your handshake to communicate confidence, warmth, and professionalism. To make sure it does, use a grip that is firm—not bone crushing—but firm. As you shake, look the other person in the eyes, and smile. Pump the other person's hand just once or twice and let go.
- *Posture.* When standing or walking, stand up straight—no slumping. When sitting, sit up straight. Good posture conveys confidence and professionalism.
- *Eye Contact.* When speaking to someone during an interview, look him or her directly in the eyes in a friendly, open manner. Also, when others are present during an interview, spread your eye contact around—make eye contact with all of them.
- *Facial Expressions.* The best facial expressions during an interview are those that convey interest, attention, and friendliness. Consequently,

you will want to make an effort to avoid facial expressions that convey anger, impatience, confusion, arrogance, or deceitfulness. Do not fall into the trap of unconsciously mimicking negative expressions you might see on the faces of those who are interviewing you. In an interview, you never know when someone might be testing you to see how you react. Consequently, regardless of what you see on the faces of others during interviews, make sure they see interest, attention, and friendliness on yours.

- *Voice Tone and Rate of Speech.* When you are nervous, as you are likely to be in an interview, your voice will sometimes reflect the fact in a higher tone or a more rapid rate of speech. This can happen without you even noticing it. Consequently, you will want to make a conscious effort to keep your voice tone and rate of speech normal during interviews. Remember, you are trying to convey an attitude of confidence and professionalism. If you feel nervous as you begin an interview, take a few silent deep breaths and let them out slowly. Then, when you speak, make a point of slowing down.

- *Gestures.* When nervous, people often engage in unconscious compensatory behaviors to make themselves feel better. These behaviors might include crossing your arms, brushing back hair, jigging keys in the pockets, rubbing your nose, and an assortment of other nervous, fidgeting behaviors. There is nothing wrong with being nervous, but you do not want to appear to lack confidence or professionalism. To avoid making distracting nervous gestures during interviews, fold your hands together in your lap and keep them there. This will also solve the problem of having shaky hands.

- *Proximity.* Proximity—in the current context—means relative distance, that is, how close or how far away you stand or sit in relation to those conducting the interview. Too close might be interpreted as being inappropriately intimate, whereas too far might be viewed as being standoffish or distant. The best distance to keep between yourself and interviewers is one arm's length. This is close enough to be involved, but not so close as to be intimate.

Ensuring Effective Interviews

The key to ensuring effective interviews is *preparation.* The better you prepare, the more likely it is that your interview will go well. Preparation for interviews is a two-step process. In the first step, you anticipate the characteristics employers are looking for and decide how you can demonstrate that you have these characteristics. In the second step, you anticipate questions interviewers might ask and plan your answers to these questions.

Anticipating the Personal Characteristics Interviewers Want

When preparing for interviews, begin by making a list of desirable personal characteristics—characteristics interviewers are likely to be looking for in you. Figure 2-6 is a checklist of personal characteristics interviewers typically look for in job applicants.

You should be prepared to demonstrate that you have these characteristics or to answer questions in ways that demonstrate the fact.

• *Honesty.* You can demonstrate honesty by answering all questions asked by interviewers in a direct and honest manner. Never make the mistake of giving an answer that might be contradicted during a reference or background check. For example, I know a student who failed to get a job he should have gotten because he answered an interviewer's question by saying he had never been fired from a job. During a reference check it came out that this young man had, in fact, been fired from a part-time job at a local grocery store only six months earlier. His firing had been the result of a minor incident that could have been easily explained and would have been understood by the interviewers. However, this student chose to cover up the incident rather than give an honest accounting of it.

Personal Characteristics Employers Look For

Interviewers look for the following personal characteristics when interviewing electronics technicians for jobs:

- Honesty

- Dependability

- Enthusiasm

- Confidence

- Teamwork

- Communication skills

- Vitality

- Perseverance

- Positive, can-do attitude

- Supervisability

FIGURE 2-6 Be prepared to demonstrate that you have these characteristics to those who interview you.

- *Dependability.* You can demonstrate dependability by showing up on time for your interview—actually you should show up 10 minutes early. In addition, if you have ever won an award for perfect attendance, punctuality, or any other aspect of dependability, be prepared to say so during interviews.

- *Enthusiasm.* If you cannot be enthusiastic about working for the employer in question, why should he or she want to hire you? Let your enthusiasm to get the job and to do a good job show in your demeanor as well as in your answers to questions asked during interviews.

- *Confidence.* You can demonstrate confidence during an interview by 1) giving straightforward answers to questions without hesitating, 2) telling interviewers you are "confident"—use the word in your answers—that you can quickly learn anything that you do not already know, and 3) asking questions of the interviewers about the job (for example, career advancement potential, what types of tasks you will be given first, how your performance will be evaluated, and so on).

- *Teamwork.* You can demonstrate your teamwork skills during interviews by 1) talking about how you have been a member of a team before—sports team, project team in school—and how you helped the team perform better, 2) explaining that you understand that team goals take precedence over individual agendas, and 3) explaining that you are willing to do what is necessary to be a contributing member of the team and to help the team achieve peak performance.

- *Communication Skills.* You can demonstrate that you have good communication skills by 1) listening carefully to all questions asked of you (remember, the most important communication skill is listening), 2) responding clearly, accurately, and succinctly to all questions, 3) looking directly at interviewers when they ask questions and when you give answers, and 4) paraphrasing and repeating back to interviewers any questions you do not understand before answering.

- *Vitality.* You can demonstrate your vitality—energy, health, pep—by telling interviewers that you are in excellent health and about anything you do that requires energy, pep, and get-up-and-go. For example, do you work while going to school? Are you a part of organizations or teams that require a lot of your time outside of class? It takes a lot of energy to go to school and do these other things at the same time. Do you have a hobby that requires a lot of energy (for example, running, biking, walking, mountain climbing, kayaking, and so on)? If so, work the fact into one of your answers during interviews.

- *Perseverance.* You can demonstrate you willingness to persevere when the work gets tough by telling interviewers of instances in your life when it was necessary to persevere. If you had to work to put yourself through school, say so. If you had to spend long hours completing your schoolwork, talk about your experiences. I know a student who won a job over several other equally qualified applicants because of how he had persevered to finish school on time after suffering serious injuries in an automobile accident. I know another student who did well in an interview because he had persevered in helping his mother support the family after his father walked out on them. If you have had to face adversity in your life, and you hung in there until you could overcome it, tell interviewers about your experiences.

- *Can-Do Attitude.* You can demonstrate that you have a positive can-do attitude by telling interviewers about instances in which you were asked to take on new and unfamiliar challenges and how you responded by saying "can-do" and got the job done. When working in electronics, you will face new and unfamiliar challenges all the time. Employers need to know you will not shrink from taking on work that is new territory for you or challenging projects that might overwhelm others. Tell interviewers about instances in which you have done these things, whether in school or in another setting.

- *Supervisability.* You can demonstrate your willingness to submit to the authority of a supervisor by telling interviewers of instances when you have done so. For example, I once served on a committee to hire a technician in which supervisability was a key selection criterion because our company had recently fired another technician for insubordination because he had refused more than once to follow appropriate directions from his supervisor. This supervisor was on the selection committee, and he asked the following question: "Are you good at following the directions of a supervisor?" The applicant responded: "I just got out of the Navy where I learned to follow all lawful orders in a positive manner. I will do the same in this job if hired." The applicant in question was well qualified, but it was this answer that made him stand out in a positive way from the other qualified applicants.

Anticipating Interview Questions

Interviewing for a job is a lot like taking a test—you do better if you know what to expect. Interview questions that catch you unaware or off guard can make you look unprepared, tentative, or ill-informed. To minimize the chances of this happening, think about the questions you are likely to be asked during an interview and how you will answer these questions. The following list provides the types of nontechnical questions most frequently asked during interviews for electronics jobs. You should also be prepared to answer technical questions relating to the job.

- Why do you want to work here?
- What do you know about our organization?

- Why do you think you are qualified for this job?
- Are you able to take directions from a supervisor in a positive manner?
- Do you find it difficult to work with people who are different from you?
- Can you disagree with others without being disagreeable?
- Did you have a job while you were in school?
- Are you willing to take on new and unfamiliar tasks?
- Are you punctual and regular in your attendance?
- Are you willing to work overtime if circumstances require it?
- Are you able to work well as a member of a team?
- Do you have difficulty trying to communicate with others?
- Do you have the energy to work long shifts when necessary?
- Where would you like to be in your career five years from now? Ten?
- If we offer you this job, when can you start?
- Do you work well without close supervision?
- Are you comfortable meeting and talking to strangers (asked of applicants who will make service calls)?

Different people will ask different questions during interviews, and no matter how well prepared you are, there may still be surprises. The key is to minimize the surprises by anticipating the types of questions you will be asked. The list of questions just provided will help, but do not base your preparation on just this list. Ask electronics technicians who are currently employed in the field what types of questions they were asked during interviews. In addition, ask anyone you know who interviews technicians to tell you the types of questions they ask.

Concluding an Interview

You should try hard to make a good *first* impression on interviewers, but it is also important to leave them with a good *last* impression. To make a good last impression, make sure that you conclude the interview in a positive manner. The following are some strategies for concluding interviews in a way that will leave interviewers with a positive last impression of you.

- If a job offer is not made on the spot, ask when they plan to make their decision.
- Get a business card from the person in charge of the hiring committee (so that you can follow up with a thank-you note).

- Thank all interviewers for their time and consideration. Look each person in the eyes, smile, and shake hands.
- Make sure the last words you say are these: "I would like to have this job, and if you give me a chance I will do an excellent job for you."

Following up After an Interview

If an interview is inconclusive—as is sometimes the case—proper followup will be important. In most cases, organizations interview several applicants before making a job offer to one of them. The typical approach is to 1) conduct all of the interviews, 2) give members of the hiring committee a couple of days to think about the various applicants, and 3) rank the applicants in priority order. The applicant who is ranked first by the majority of the members of the hiring committee is then contacted and offered the job. The entire process can take from a few days to a couple of weeks.

Consequently, at the end of an interview, you may not yet know if the job is yours. This is one of the reasons you asked for a business card from the person in charge of the hiring committee at the conclusion of the interview. Within 24 hours, you should follow up with this person by e-mail. In days past, this type of followup was accomplished by telephone or letter, but e-mail has quickly become the method of choice because people at both ends of the communication tend to prefer the immediacy and convenience.

Reasons for Sending a Follow-up Note

Sending an e-mail note to the person in charge of the hiring committee can give you an advantage over other applicants if it is done properly. A well-written e-mail note will achieve the following:

- Makes getting in touch with you as easy as clicking Reply.
- Shows interviewers that you really want the job.
- Gives you one more chance to make a positive impression on interviewers by showing that you are considerate and that you follow up on important tasks.

Writing Your Follow-up Note

Do not just sit down at your computer and dash off a quick followup note. This note, though brief, could turn out to be just as important as your interview. Preparation is an important factor here. Begin by typing up a rough draft and saving it. Print the draft and spend some time editing and revising it. Keep the following strategies in mind as you edit and revise your rough draft:

Dear Mr. _____:

Thank you for the opportunity to interview for an Electronics Technician position yesterday. Having discussed the position with you and your colleagues, I am convinced that this is the right job for me and that I am the right technician for this job. I hope you will give me the opportunity to show that I can do an excellent job for your company.

I can begin work immediately. Thanks again for the interview. I look forward to hearing from you.

Respectfully,

Mary Arballo

FIGURE 2-7 It is important to follow up with a positive e-mail note within 24 hours of the interview.

- Keep your note brief, upbeat, and positive.
- Say that it was a pleasure to meet you, and insert the names of all interviewers. For men, use the title "Mr." and for women use "Ms."
- Make the points you think are important using words and phrases that reinforce the personal characteristics interviewers look for in applicants (for example, I will be "dependable," "enthusiastic, "a team player," and so on).
- Make sure the signature block on your e-mail note contains your name, address, and telephone numbers (including cellular). In other words, make it easy for employers to contact you. Figure 2-7 is an example of a well-written follow-up note.

Review Questions

1. Describe the needs you have relating to your job as an electronics technician.
2. What needs does the employer have that you will have to satisfy as an electronics technician?
3. List and explain the characteristics of a well-developed resume.
4. Compare and contrast the three types of resumes you might use when seeking a job.
5. What are three purposes served by a cover letter?

6. Explain the difference between primary references and secondary references.
7. Describe how you would use the Internet to help locate job openings.
8. What are the personal characteristics employers are looking for in applicants during interviews?
9. Explain how you can send the right nonverbal signals during an interview.
10. Explain how to conclude an interview on a positive note and how to follow up after an interview.

Discussion Questions

1. Defend or refute the following statement: "As long as I make enough money, I don't care whether I like my job or not."
2. Discuss with classmates how you would go about putting together a comprehensive job-seeking portfolio. What would you say to a fellow student who does not think he needs a portfolio?
3. Defend or refute the following statement: "If my electronics skills are good enough, I'll get hired. Nothing else matters to employers."
4. Discuss with classmates how you would prepare yourself for an interview. Call on any experiences you may have had interviewing—even for part-time jobs.
5. Discuss what you would do to make a good *first* impression during an interview and a good *last* impression. Why do you think it is so difficult to overcome a bad first impression?

Application Assignments

1. Develop a resume you can use when applying for jobs after you complete your electronics program. Choose the type that best fits your level of experience.
2. Develop a sample cover letter to put in your job-seeking portfolio.
3. Develop a list of primary personal, work-related, and school-related references, and secure permission to use their names during your job search.
4. Develop at least one secondary reference for each primary reference identified in Assignment 3.
5. Go to several Internet sources, and identify as many jobs in your field as you can. Now add to the list using newspapers and employment agencies.

6. Using the Internet as a resource, identify job fairs that will occur within a reasonable distance from your home. If possible, attend at least one job fair while you are still in school, and report back to your class concerning what you learned.

7. Work with your classmates to organize mock interviews in which you ask each other interview questions and practice your answers to those questions. Ask another classmate to summarize all of the nonverbal signals you send while responding to interview questions and to share those summaries with you. If possible, record the interviews on DVD or videotape.

8. Develop an example of a follow-up note to an employer you could use after an interview. Put the sample note in your job-seeking portfolio for using later.

CHAPTER 3

BUILD YOUR CAREER ON A FOUNDATION OF INTEGRITY

LEARNING OBJECTIVES

Upon completion of this chapter, you should be able to do the following:

- *Define the term* integrity.
- *Explain what makes maintaining your integrity such a challenge in the modern workplace.*
- *Explain why cheating has become so common in our society.*
- *Describe the typical long-term results of dishonesty.*
- *List the personal characteristics of people with integrity.*
- *Explain why integrity is so important in building a successful career.*
- *Explain the relationship between integrity and ethics.*
- *Demonstrate how to make ethical decisions.*

In the long run, no characteristic is more important for people who want to build a successful career than integrity. Notice that I said, "In the long run. . . ." Over the span of your career, it will sometimes appear that the best choice in a given situation, at least in the short run, is the half-truth, shortcut, or unethical decision. However, in the long run, integrity almost always wins out over short-term gains won by telling half-truths, taking shortcuts, or making unethical decisions. The term *ethics* concerns the practical application of moral principles and values. Making the "ethical choice" means doing the right thing within a moral framework that is built on honesty and integrity.

Often, the most difficult choice to make is the ethical choice—the choice of a person with integrity. There are several reasons for this, but the most common are self-interest and the natural human desire for immediate tangible results. As a human being, we tend to want what we want (self-interest), and we want it now (immediate gratification). Often, it appears that the fastest way to get what you want and get it right now is to take an ill-advised shortcut, to tell someone a half truth, lie, cheat, or make a decision you know is unethical and hope you won't get caught.

For example, you probably know of a fellow student who has cheated on a test in order to pass. In the short-term, the immediate payoff for cheating is a passing grade on the test, but what is going to happen in the long run when he needs the information he did not know on the test in order to do his job? Further, what is going to happen when he begins to cheat on a regular basis rather than studying because he got away with it once or twice? The fact is that he will eventually get caught. Even if he doesn't get caught while in school, his cheating habit will eventually catch up with him—it always does—and when it does, the consequences can be dire.

The ethical actions of people with integrity rarely pay immediate dividends. Typically, the tangible rewards of integrity come in the long run, not the short term or immediate future. The internal, intangible reward of making ethical choices is immediate; it is a clear conscience. But the external, tangible rewards of making ethical choices—rewards such as recognition, wage increases, promotions, and bonuses—often are not. However, never lose sight of the fact that the tangible rewards of integrity—when they do come—are yours to keep. You can rest easy with a clear conscience knowing that rewards won with integrity will not be taken away from you by scandal, ethics violations, or broken laws. Remember, you never have to worry about "getting caught" for doing what is true, right, and honest.

Integrity Defined

Integrity means living your life and practicing your profession in a way that is consistent with the core moral values of honesty and truthfulness. A person with integrity always tries to do what is right in a given situation rather than what is self-serving or expedient. People with integrity are guided by honesty and the truth rather than self-interest and the potential for personal gain. Living a life of integrity requires that your beliefs and your behavior be consistent, that your actions match your moral values, and that your moral values center around honesty and truthfulness.

If you have integrity, you will tell the truth, take the proper action, or make the right decision even when it does not appear to be in your best short-term interests to do so. Your integrity or lack of it will show up in the choices you make, the actions you take, and the behavior you exhibit on a daily basis. Over the past 40 years, I have seen countless individuals get ahead temporarily in their careers by taking unethical shortcuts, lying, cheating, taking advantage of others, or claiming credit for the work of others only to have their lives fall apart later when the truth finally rose to the surface—as it inevitably does. One of the most seemingly successful people I ever worked with is still serving time in a federal penitentiary to pay for the unethical choices he made in order to get ahead.

Why Is Working with Integrity Such a Challenge?

Why is working with integrity such a challenge for people? The easy answer is human nature. As people, we tend to operate on the basis of *perceived self-interest*. During the course of discussing some idea, you have probably heard someone say, "What's in it for me?" The person who made this comment was concerned with what he perceived as his self-interest. People often look at situations from the perspective of what's in it for them because it is just human nature to look out for number one.

Unfortunately, there will be many times at work and in life when your *perceived self-interest* will make it appear that an unethical choice is the best thing to do for yourself—at least in the short run. This is often where people get in trouble. So many times in the workplace, people make unethical choices because they focus on the short-term outcome without considering the long-term consequences of their choices. They tell themselves: "I probably won't get caught—I'll worry about the consequences later."

Perceived personal-interest is a broad concept that covers several different factors, the most common of which are *greed, impatience, ego, fear, expedience,* and *ambition.* For example, greed might cause a person to claim more hours on his timesheet than she actually worked. In the short run, this would bring her more money in her paycheck, but in the long run, it could get her fired or worse. Impatience might cause a person to take a shortcut when performing maintenance on an important piece of equipment. In the short run this might let him get the job done sooner so he can go home on time, but in the long run, it might cause the equipment in question to malfunction. When it is determined that the malfunction was caused by maintenance shortcuts, the technician who took those short cuts might lose his job or be required to pay for the damage.

Ego might cause a person to exaggerate his abilities to get a promotion and a raise. In the short run, this would bring him more money and status, but in the long run, it will cause problems when it becomes apparent that he cannot really do the job that came with the promotion. Fear of retribution might cause a technician to look the other way and ignore the issue when he knows that a teammate is cheating on his expense voucher or doing something else that is unethical or illegal. In the short run, this will ensure that the technician in question is protected from the anger of his teammate, but in the long run, he could be viewed as an accomplice to the fraud when it is learned that he knew but failed to do anything.

Expedience might cause a person to cut corners on work projects to get more of them done in a day than her teammates—thereby making it appear that she is more productive. In the short run, this might lead to such rewards as a cash bonus, raise, or some type of recognition award. But in the long run, when the work projects in question begin to fail quality tests or break down when customers try to use them, the technician who took the shortcuts will be disciplined and might even lose his job. Misguided ambition might cause a technician to begin sabotaging the work of a teammate in order to outperform him and win the competition for a promotion. In the short run this might win the technician in question the promotion, but in the long run, when it is determined that he sabotaged the work of a teammate, he will lose the promotion and possibly his job. If you are thinking that you pay the consequences of unethical behavior only when you get caught, and that you might not get caught, remember that the jails and prisons in this country are full of people who thought they wouldn't get caught.

From these examples, you can see that what appears to serve your self-interest in the short run may actually hurt you in the long run. Practicing your profession with integrity often means having to overcome the temptations of misguided short-term self-interest. To succeed in the long run, you have to be willing to put aside the immediate gratification that can sometimes be gained by making unethical choices and work toward the longer-term rewards that come from doing the right thing.

Why Cheating Has Become So Commonplace

Is cheating something everyone does? Have we become a society that no longer cares if people cheat? Do we actually condone or even encourage cheating? According to author David Callahan, America has developed a "cheating culture."[1] Callahan cites numerous examples that support his contention that cheating in America has become a widely accepted practice that no longer generates the shame it once did in people who are caught doing it.

Callahan cites such examples of endemic cheating as 1) psychiatrists who accept payments from wealthy parents to falsely diagnose high-school students as having learning disabilities so they will be given more time to take their SAT tests for college admission; 2) a corporate CEO who is caught claiming to have an MBA degree when, in fact, he never even went to graduate school; 3) a businesswoman who keeps the receipt for the entire cost of a meal that she and several other colleagues divided up and then claims the entire cost on her expense account for reimbursement; 4) thousands of people who use technology to illegally download music from the Internet, thereby robbing the artists who created the music of revenue that is rightly theirs; 5) a newspaper reporter who fabricates quotes and makes up stories rather than doing the necessary research for his column; and 6) corporate executives who admit that they regularly cheat in business-related golf matches.[2]

Callahan's concern is not just that people cheat, but that cheating seems to be so widely accepted. It's as if people no longer see anything wrong with cheating. Callahan gives the following reasons for society's widespread acceptance of cheating:[3]

- *New pressures.* Now that competition in the marketplace has become global, the pressure on people to perform at peak levels has become so intense that it can overwhelm their sense of right and wrong. Some people are so desperate to perform at a competitive level that they see cheating as the only way they can do it.

- *The payoff can be so attractive.* Some people believe that if the payoff is big enough, they don't care if they have to cheat in order to win it. In other words, some people want to win the promotion, raise, recognition, award, or competition so badly, they adopt an attitude that the end justifies the means. The end is whatever reward they want, and cheating becomes the means for achieving it.

- *No apparent downside.* The temptation to cheat has increased in proportion to the decrease in the ability or willingness of organizations, agencies, authority figures, and the legal system to do anything about it. In other words, so often it seems that people cheat and get away with. This leads others to think: "If they can cheat and get away with it, why shouldn't I?"

- *So many bad examples to follow.* Students who, during school, see stories in the news about corporate executives, elected officials, or other prominent figures who have benefited in some major way from cheating, might think, "If this is how prominent people get ahead, why shouldn't I follow their example."

Long-Term Results of Dishonesty and Cheating

The previous section summarized some factors that make cheating and dishonesty appear so attractive to some people. And there is no question that people often appear to benefit in the short run from cheating and other types of dishonest behavior. However, in the long run, these same people almost always end up paying a heavy price for their unethical choices. Consequently, when you see people who appear to benefit from cheating, remind yourself that the story is not over with yet, and the end of the story might not be as bright for these people as the beginning appears to be at the moment.

The following example illustrates a hard truth about practicing your profession without integrity. Invariably when people let misguided self-interest cause them to behave in an unethical manner, the truth eventually comes out, and when it does, the long-term negative consequences typically outweigh the earlier short-term benefits.

John and Joe grew up together, graduated from high school in the same class, attended the same community college, completed the same electronics program, made the same grades, and graduated on the same day. On the surface, John and Joe looked like carbon copies of each other. Their resumes were interchangeable. However, as it turned out, John and Joe were very different types of people.

After graduation, John and Joe went to work as electronics technicians at different companies in the same town. They began their careers at the same level, but within just a couple of years, it became clear that John was outrunning his friend as they competed to climb the career ladder. John was being promoted faster and making more money than Joe; a fact that was obvious to friends who knew them both. John's house and car were big, new, and expensive, while Joe's little apartment and old car paled by comparison. At first their mutual friends attributed John's evident success to superior performance on the job. They assumed that John was working harder and smarter than Joe and his other contemporaries. They had seen John work hard and work smart in school and knew what he was capable of when he set a goal for himself. Then, one day while the two friends were having lunch, John brought up the subject of Joe's comparatively slow career advancement.

The lunch discussion turned out to be a real eye opener for Joe. He learned that John was more talented at self-promotion, office politics, backstabbing, cutting corners, stretching the truth, and collecting unauthorized "extra fees on the side" from customers (fees that he never reported). It was these unethical actions, not his talent and hard work, that were pushing John up the career ladder so fast. When Joe questioned the ethics of his behavior, John just laughed and told Joe he was naïve. John's defense was that "Everybody does it, so why shouldn't I?"

After this disturbing discussion, the two friends drifted apart. Joe was no longer comfortable being around John. As Joe continued to slowly but surely advance in his career, he kept tabs on his old friend's continued rapid progress. By the time Joe was a team leader, John was the director of his company's electronics service department. The gap between their circumstances became so broad that Joe began to doubt his own value system. He began to think that perhaps John was right. Maybe integrity was just something for naïve people who were suckers.

While Joe was struggling with this crisis of conscience, he opened his newspaper one morning and happened to glance at the "police "blotter" column. To his astonishment, Joe found himself looking at a photograph of a very unhappy John. The caption under the photograph read: "Local man arrested for fraud." John's lack of integrity had finally caught up with him. Over the years, John had made a practice of charging his company's customers extra fees that were paid in cash in order to have their work moved to a higher priority on his work list. In other words, John would tell a customer: "For some unreported cash on the side, you can have your equipment serviced and repaired today instead of waiting your turn on the service list."

John's scheme worked well for a while; that is, until he got greedy. Eventually his extra fees became so exorbitant that customers began to complain, and the company started an investigation. The investigation revealed that John's fraudulent scheme involved so much money that not only did he have to be fired, he had to be turned over to the police. During the trial, John complicated his problems by trying to bribe a juror to rule that he was not guilty. When this was discovered, it only added to the charges John faced in court and ensured that John would spend a long time in prison.

Like so many people who cheat in order to win, John became accustomed to the practice and, before long, it was the only way he knew how to win—hence his attempt to bribe a juror. Not all people who practice their professions without integrity go to jail, but most eventually pay a price for their lack of honesty. And often the price is high—so high that it outweighs the short-term benefits they gained from behaving unethically.

Personal Characteristics of People with Integrity

People of integrity behave according to the same standards in every situation. Whether the boss is watching or not, they do what is right. This is a personal characteristic that is common to people of integrity.

Checklist of the
Personal Characteristics of People with Integrity

- Consistent

- Values-driven

- Predictable

- Fair and equitable

- Substance-oriented

- Selfless

FIGURE 3-1 People with integrity have these characteristics.

What follows are some other personal characteristics common to people of integrity:

- *Consistency.* People of integrity are not actors or hypocrites. They don't put on a show for others and then behave differently when they aren't being watched. They are who they are. People of integrity don't change their values or behavior to match ever-changing social trends, go along with the crowd, or submit to peer pressure when making decisions. They consistently do what is right in the situation in question.

- *Guided by their values.* People of integrity have a strong value system that is so much a part of them that the person and values are insepara- ble. All aspects of their lives are guided by their values, which include honesty and integrity. Because of their values, people of integrity admit it when they make mistakes or are responsible for the errors of others. In addition, they behave as if they are always being observed by the ever-present eye of their moral compass—which they are.

- *Predictable.* Because their behavior is consistently guided by their per- sonal value system, people of integrity are predictable. You don't ever have to wonder what a person of integrity will do in a given situation. They are predictable in that they will do the right thing; even if it is not the popular, convenient, or expeditious thing to do. They are the same person when out of town that they are when at home. Because they are predictable, they can be depended on to keep their word.

- *Fair and equitable.* Within the limits of morality and ethics, people of integrity make their decisions based on what is best for their teams, organizations, or the greater good, rather than self interest. Where less ethical people will try to cloud the issues and talk about "gray areas," people of integrity use their moral compass to see through the fog that can be generated by misguided self-interest, theirs and that of others. Because this is the approach they take, people of integrity can be counted on to be fair and equitable in their dealings with others. They don't play favorites, and they don't treat one teammate one way and another teammate in a different way. In all cases, whether the team-mate is a friend or just another employee, people of integrity will treat him or her fairly and equitably.

- *Put substance over image.* We have become a society that is obsessed with image. Politicians often concern themselves more with how they look on television than how informed their views and opinions are. Although you should certainly be concerned about your image, don't ever make the mistake of thinking that image is more important than substance. People of integrity know that they must be able to do the job, not just look like they can do the job.

- *Selfless.* People of integrity think of the team first and their individual needs second. You cannot be a self-serving person when dealing with others and also be a person of integrity. Technicians who are ultimately the most successful are the ones who are good stewards of the resources for which they are responsible—human, financial, and physical. Self-lessness means that you think of being a good steward toward these resources before thinking of yourself. Selfless people care about the greater good of the team and the organization.

Importance of Integrity in Building a Career

To succeed as an electronics technician, you must become a leader among your teammates. To lead, you must have people who are willing to follow you. This is why integrity is so important to those who want to be leaders. A leader can lead only if people will follow, and people will follow only if they believe in, trust, and respect the leader. Without integrity, people will not believe in, trust, or respect you. Here are some reasons why integrity is so important to career success:

- *Integrity builds trust.* Most of your work in electronics will be done in teams. Team members must be able to count on each other; they must be able to depend on each other. In other words, they must trust each other. When team members do not trust each other, teamwork falls apart. On the other hand, people who can be depended on and trusted are more likely to get the cooperation and support of teammates.

- *Integrity leads to influence.* You cannot just tell people what to do. Even when you are the boss and in a position to give orders, there are limits.

For example, the boss can order you to do something, but he cannot force you to put your whole heart into it. Consequently, it is important for people in positions of authority to have influence with their direct reports. Leaders who can influence their team members in a positive way are able to persuade them to make a real and total commitment to achieving the team's goals. You cannot have this kind of influence with team members unless they know you have integrity.

- *Integrity establishes high standards.* People with integrity consistently set a positive example of doing the right thing in every situation, even when doing the right thing is the most difficult option available to them. By their positive examples, people with integrity set high standards for their teammates, and having high standards is essential in a competitive environment.

- *Integrity establishes a foundation of substance rather than just an image.* While image can be an important factor in success, image without substance will eventually lead to failure. People of integrity—although they understand the importance of a positive professional image—put substance first and image second. They know that the foundation of their credibility is built on integrity. The good news is that, in the long run, a reputation for integrity will go a long way toward giving you a positive professional image.

- *Integrity builds credibility.* To influence your teammates in a positive way, you must have credibility with them. Credibility is what you have when others believe in and respect you, and it is a critical characteristic for those who are building a career. To establish credibility, there must first be trust. If you want to establish credibility, first establish trust.

Integrity and Ethics

Ethics is the practical application of morality. It answers in a practical, everyday sense the question: "If I believe this, how then should I live?" It follows that *ethical behavior* means doing the right thing within a given value system. A value system is a set fundamental beliefs about what is right and wrong. On the job, that value system should be the corporate values of your employer. This is why an ethical person should not work for an unethical employer. If your personal values don't match the corporate values of your employer, you will constantly find yourself in potentially compromising situations.

To behave ethically, you must have an established framework—a set of values that guide your behavior, decisions, and choices. To succeed in the end, people must set a consistent and positive example of ethical behavior. To set such an example, they must be willing to live and practice their profession in accordance with an appropriate set of values. These values must begin with honesty and integrity.

Electronics technicians are often faced with ethical dilemmas in which they must balance their self interests with the needs of the organization, make decisions that can affect the organization's profitability in both the short and long run, balance their personal responsibilities to family and work, and withstand both overt and covert pressure to cut corners in the name of short-term profitability. I know an electronics technician who eventually had to change jobs because his employer was constantly badgering him to charge customers for work that was not done or to do work that was unnecessary. He viewed these practices as cheating. Consequently, he tried for more than a year to convince his boss to do what was right when billing customers. When he persisted in refusing to cheat customers, his boss began to harass him in a variety of ways. Finally, when this technician had taken all of the harassment he intended to take, he sought and won a job at another company. But that wasn't all he did. In addition to leaving his unethical employer, he allowed his loyal customers to follow him to the new company.

Making decisions that have high ethical content can cause people to undergo what is commonly known as *soul searching.* Soul searching, as typified by the ethical technician from the previous paragraphs, amounts to weighing what you truly believe is right against the various pressures you feel when making decisions, while also factoring in the potential personal consequences of deciding one way or the other.

For example, the honest technician from the previous paragraphs went through some difficult times and some real soul searching while working for an unethical boss who constantly pressured him to cheat. At one point his boss had threatened to fire him and give him a bad recommendation if he tried to find another job in the field of electronics. With a family to support, the technician knew he had to take this threat seriously, and he spent many troubled hours worrying about it. For a while he even considered giving in to the pressure and just adding the unethical extra costs to his customers' bills to satisfy his boss and to give himself some economic security. He was especially concerned about losing his health insurance because, at the time, his wife was pregnant, and he really needed good medical coverage.

He finally decided that cheating is cheating, even when your wife is pregnant. He also decided that the best solution to his problems was to find another job, which he did without using his unethical boss as a reference. In fact, he asked the employer who eventually hired him to refrain from calling his boss out of fear of retribution—a practice that is common when experienced technicians look for a new job. With his new employer, this ethical electronics technician quickly established himself as an outstanding team member and, as a result, enjoyed a very successful career.

Making difficult decisions is one of the most frequently faced dilemmas of people who are committed to practicing their profession with integrity. This is because in order to make a decision that is *right* as opposed to *popular,* you must have the moral courage to stand up to the pressure that will inevitably be put on you. For example, think back to the ethical technician from the preceding paragraphs. Not only was he pressured by his boss to unethically alter the bills of customers, he was even pressured by his wife to give in, at least temporarily, because she was afraid he would lose his job and, in turn, his health insurance while she was pregnant. That is a lot of pressure to stand up to.

What makes living ethically even more difficult is that people who want you to do otherwise will not just sit back and congratulate you for being an honest person. On the contrary, they typically try to find ways to harass and discredit you. Consequently, you should never be so naïve as to expect those opposed to your ethical behavior to just sit back and willingly accept it. If the person who is unhappy with your decision is a customer, he might complain to your boss about the quality of your work, even though he knows your work is excellent. You might then find your job security threatened because your ethically correct decision caused your company to lose an important customer.

If the unhappy person is your supervisor, doing the right thing might mean that he retaliates by giving you low ratings on a performance appraisal. This, in turn, will threaten your chances for promotions and raises. If the unhappy person is a teammate, your ethical choice might cause him to become angry and start circulating untrue but damaging rumors about you. There is almost always a short-term price to pay for behaving ethically when others want you to do otherwise. However, remember this unalterable fact of life in the modern workplace. When it comes to ethics and integrity, it is not who wins in the short term that matters—it is who wins in the long run.

For example, recall the ethical electronics technician who refused to add improper charges to the bills of his customers. About a year after he left the company that was regularly cheating its customers, a scandal erupted when a technician who had participated in the unethical billing practices himself called a big and important customer and became a whistleblower. Angered because his boss refused to give him a raise, this technician called the customer and told him everything. The customer, in turn, contacted his attorney. An investigation ensued, and the unethical company eventually went out of business from the bad publicity. Even though as part of the legal settlement of the case the unethical company paid most of its customers back what it had overcharged them—an amount that was well over a million dollars—customers no longer trusted the company. Therefore, they refused to do business with the company, and it went bankrupt. All of the electronics technicians who had given in to the pressure to unethically alter the bills of customers lost their jobs. What made the case even sadder was that these technicians—who were not bad people, just people who let themselves be pressured into making bad choices—found it difficult to get new jobs in their field because of the "stain" of unethical behavior on their records.

Making Ethical Choices

Sometimes the ethical choice in a given situation is obvious. But there will be times when you simply will not know what the right choice in a given situation is. Sometimes the indecision is caused by a desire to rationalize a decision you know in your heart is unethical, but that you want to make anyway for reasons of self-interest (greed, fear, ego, and so on). Such situations are just a matter of doing what you know is right in spite of your desire to do otherwise and are not the subject of this section. However, there will be times when even the most ethical people will not know for certain what the right thing to do is in a given situation.

For example, I can remember when as a supervisor, a technician in my department cheated on his timecard. He was recording more overtime than he actually worked. On the surface, this probably sounds like a black and white case of cheating—and it was. The problem that clouded the issue is that he was using the extra money to help take care of his father who was confined to a rest home for the elderly. If this person had been using the money for his own personal benefit, I would have known immediately what to do: either have fired him or require him to return the unearned wages and submit to some type of discipline. But because he was trying to be a good son and take care of an elderly father who could no longer take care of himself, I was stumped concerning what I should do.

Fortunately, at the time, I was going to college at night and had a wise professor I could approach for advice. This professor asked me to consider how I would feel about my handling of this situation if all the details about it were printed on the front page of my local newspaper to be read by my family, friends, colleagues, and teammates? Then he told me I should never make the mistake of letting extenuating circumstances cloud my judgment. He said "Where is it written that having a father in a rest home makes cheating on your timecard right?"

This professor said "You need to separate cheating on the time card and having a father in a rest home in your thinking. You are letting the extenuating circumstance cloud your judgment concerning the employee's unethical behavior." In other words, I knew that cheating on a timecard was wrong, so I should take the appropriate action to stop the unethical practice immediately and apply any discipline that might be deemed appropriate. But, I also knew that having a father who could no longer care for himself lying in a rest home was a difficult and expensive problem to have to deal with. The professor suggested that I might want to find some way other than allowing an employee to cheat on his timecard to help him provide for his father.

Based on this conversation, I put an immediate stop to the timecard cheating. Our company magnanimously did not require the employee in question to pay back any of the extra wages he had been paid. In addition, a fundraiser was held that eventually spread to other organizations outside of our company, and sufficient money was raised to make a substantial contribution toward paying the cost of the rest home's services. In fact, more funds were raised in this way than the technician could ever have generated by cheating on his timecard.

When you are faced with an ethical dilemma and do not know what action to take, remember the lessons of this case:

- Ask yourself how you would feel about your decision, choice, or actions if the full story was printed on the front page of your hometown newspaper.
- Don't let extenuating circumstances cloud your judgment. Deal with what you know is right, and find other ethical ways to deal with the extenuating circumstances.

Review Questions

1. Define the term *integrity.*
2. Explain why it can be so difficult to practice your profession with integrity in the modern workplace.
3. Explain the reasons why cheating seems to have become so commonplace in today's society.
4. Typically, what are the long-term consequences of dishonesty?
5. List and explain the characteristics that are common to people of integrity.
6. Why is integrity so important for those who are trying to build a successful career in electronics?
7. What is meant by *ethics,* and how does the concept relate to integrity?
8. Explain how you would make an ethical decision when you are not sure what the right thing to do is.

Discussion Questions

1. Discuss the following topic with classmates: What are some situations you have found yourself in where maintaining your integrity was difficult? Discuss why it was so difficult.
2. Defend or refute the following statement: "Cheating is so common that people no longer think anything is wrong with it." Give examples to support your opinion.

3. Think back in your life and remember any situation you can in which someone paid a heavy price for unethical behavior. Discuss what happened with your classmates.

4. Discuss the following ethical situations with your classmates. What do you think is the right think to do in each case and why?

   a. Thousand of children across the world die of hunger and disease every day. If for reasons that cannot be explained you could stop this unfortunate reality by killing just one person yourself, would you do it?

   b. It is late at night, and you stop by the grocery store to make a purchase. While getting out of your car, you open the door and put a small dent in the door of the car next to yours. It is a brand-new, expensive car. Would you simply drive away and forget it, or try to find the owner and explain what happened?

   c. You are about to take the last test you need to take to complete your electronics program and graduate. You have not prepared and are concerned that you won't pass. You notice that you can easily see the paper of the person who sits in front of you, and he is really smart. If you knew the teacher would not catch you, would you cheat?

Application Assignments

1. Do some Internet research into the issue of "business ethics," and identify several situations in which someone has ruined his career by making unethical choices. Report these examples to your class.

2. One of the most common sources of unethical behavior on the job is fraudulent workers' compensation claims field by people who claim to have been injured on the job but either are not really injured or are injured but not by anything that occurred on the job. In both cases, they are collecting workers' compensation insurance payments fraudulently. Do some Internet research on the topic of "Workers' Compensation Fraud," and share the results with your class.

3. Identify someone who has worked—preferably in electronics—for more than 10 years. Ask this person to tell you about the most difficult decision he or she has had to make in terms of ethics. Share what you learn with your class.

4. Talking with the same person identified in Assignment 3, ask this person to tell you about ethical problems he or she has seen occur in the workplace and the effect they had on those involved as well as others who were not involved.

Endnotes

1. Callahan, D., *The Cheating Culture* (Harcourt, Inc., Orlando, Florida: 2004), viii.
2. Ibid, 8–12.
3. Maxwell, J. C. *Developing the Leader Within You* (Thomas Nelson, Inc., Nashville, Tennessee: 1993), 38–44.

CHAPTER 4

UNDERSTAND YOUR EMPLOYER'S "BIG PICTURE" AND WHERE YOU FIT INTO IT

LEARNING OBJECTIVES

Upon completion of this chapter, you should be able to do the following:

- *Explain the concept of your employer's "big picture" as it relates to you.*

- *Describe how you can use your employer's organizational chart to understand the big picture as well as how and/or where you fit into it.*

- *Explain how your team charter can help you understand the big picture as well as how and/or where you fit into it.*

- *Explain how your job description can help you understand the big picture as well as how and/or where you fit into it.*

- *Describe how you can use your employer's performance appraisal process to understand the big picture as well as how and/or where you fit into it.*

People who work in an organization are like pieces in a complex jigsaw puzzle; there are lots of pieces, and every piece has an important role to play in making the puzzle come together. What is especially important is to know exactly *where* and *how* each piece fits into the overall puzzle. Where and how represent two different questions. Think of your employer's organization as a jigsaw puzzle and yourself as a very important piece in that puzzle. Your employer has a purpose for being in business, and you are an important part of helping your employer fulfill that purpose. Without you fitting in at just the right place (*where* you fit in) and in the right way (*how* you fit in), the puzzle cannot come together successfully. But if all employees know where and how they fit into their employer's organization—in other words, if they know the role they play in helping their employer succeed—the puzzle will come together just right. When this happens, your employer wins and you win.

Consequently, knowing where and how you fit into your employer's organization can be important to your success. It helps you understand what your employer needs from you, and providing what your employer needs is one of the ways you will succeed as an electronic technician. This chapter explains several "tools" you can use to develop an accurate understanding of your employer's big picture as well as where and how you fit into it.

Your Employer's Big Picture

Every organization that might employ you after you complete your electronics studies has a big picture. This big picture is a composite of the organization's purpose—why it exists—and its aspirations—what it hopes to achieve. You and every other employee in the organization are part of this big picture. Several tools are available (at least they should be) to help you see where and how you fit into your employer's organization. These tools include your employer's organizational chart, team charters, job descriptions, and performance appraisal instruments. To do your part to help the organization fulfill its purpose and achieve its aspirations, you must know how and where you fit into the big picture.

Helpful Big Picture Tools

You will need to know how and where you fit into your employer's organization to succeed. Figure 4-1 is a list of tools that will help you understand both where and how you fit into an organization.

Where you fit into an organization concerns your physical place (for example, What department is your team in? Whom do you report to? Whom does your supervisor report to?). *How* you fit into an organization

---

**Where and How Do I Fit In?
A Checklist of
Helpful Big-Picture Tools**

_____   **Organizational Chart**

_____   **Team Charter**

_____   **Job Description**

_____   **Performance Appraisal Instrument**

FIGURE 4-1 These tools can help you determine how and where you fit into your employer's big picture.

concerns what is expected of you to be an accepted and contributing member of the organization. In the remaining sections of this chapter, you will learn how the big-picture tools in Figure 4-1 can be used to determine both how and where you fit in to your employer's organization.

Organizational Chart as a Big Picture Tool

Most organizations have an organizational chart that shows how all functional units fit together, where each unit is in the organization, and who reports to whom in the organization. Figure 4-2 is an organizational chart for the hypothetical company, XYZ, Inc.

This organizational chart shows that XYZ, Inc. has four main divisions: marketing/sales, accounting/operations, engineering, and manufacturing. A vice-president manages each division and reports to XYZ's president and chief executive officer. As an electronics technician, you would fit into either the manufacturing or engineering divisions depending on the job you are hired to do.

Figure 4-3 is a division-level organizational chart for XYZ's Manufacturing division. It shows that the Manufacturing division is divided into three departments: Production, Installation, and Service. A director manages each department and reports to the vice-president of the Manufacturing division.

Figure 4-3 also shows that each department in XYZ's Manufacturing division—Production, Installation, and Service—is divided into teams. Each team has a team leader—the supervisor—as well as a number of electronics technicians (sizes of the teams vary).

FIGURE 4-2 An organizational chart is an excellent big-picture tool.

Assume you have been hired by XYZ, Inc. to work as a service technician troubleshooting, maintaining, and repairing equipment produced and installed by other teams at XYZ. Using the organizational charts in Figure 4-2 and 4-3, you could answer the question, "Where do I fit into XYZ, Inc.?" as follows:

- You will be part of XYZ's Manufacturing division, and your executive-level manager will be the company's vice-president for manufacturing.

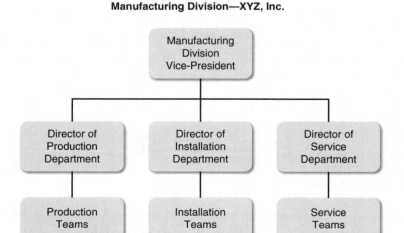

FIGURE 4-3 This organizational chart shows all of the components of just one of the four divisions of XYZ, Inc.

- Within the Manufacturing division, you will be part of the Service department, and your division-level manager will be the director of the Service department.
- Within the Service department, you will be part of one of the service teams, and your supervisor will be the leader of that team.

As you can see from this section, organizational charts are excellent tools for helping you see *where* you fit into an organization. They do not help with the issue of *how* you fit in. Other tools will answer that question.

Team Charter as a Big-Picture Tool

Work in organizations is typically done in teams. To make sure that teams know the role they play in the larger organization, some employers develop team charters (all should develop them). A *team charter* is a brief document containing the team's name, mission statement, and guidelines for how team members are expected to interact with each other and perform their jobs. A team charter shows *how* the team and individual team members fit into the larger organization. Figure 4-4 is a team charter for the service teams in the Service Department of XYZ, Inc. (as shown in the organizational chart in Figure 4-3).

Assume that you are an electronics technician on one of the service teams at XYZ, Inc. The team's mission statement explains a lot about how you and the team fit into the company's big picture. To do this, the team's mission statement must tie directly into the company's mission statement. Both mission statements are shown here to illustrate this point:

XYZ Inc.'s Mission Statement

XYZ, Inc. designs, manufactures, installs, and services electronic control systems for air conditioning and heating systems used in commercial and industrial applications in the United States and Canada.

Service Teams' Mission Statement

Service teams at XYZ, Inc. are responsible for providing maintenance, troubleshooting, and repair services on-site in the facilities of customers in the United States and Canada.

By comparing these two mission statements, it is easy to see how service teams fit into XYZ's big picture. The company has four major functions: design, manufacture, installation, and service. Clearly, the service teams fit into XYZ's big picture by fulfilling one of the company's main functions—service. The team charter also shows that to fit in at

---

Team Charter
SERVICE TEAMS
XYX, Inc. Production Department

Team Mission Statement

Service teams are responsible for providing maintenance, troubleshooting, and repair services on electronic control systems produced and installed by XYZ, Inc. All service is provided on-site in the facilities of customers.

Performance/Interaction Expectations

All members of a service team at XYZ, Inc. are expected to meet the following expectations concerning performance and interaction:

- Provide high-quality maintenance, troubleshooting, and repair services on every contract every time.

- Do what is necessary to continually improve performance—yours and the team's.

- Treat all customers, team members, supervisors, and managers with respect, patience, and a positive, friendly attitude.

- Do what is necessary to prevent/resolve conflict in the team (for example, when you disagree, don't be disagreeable).

FIGURE 4-4 A team charter is an excellent big-picture tool.

XYZ, Inc., service technicians will have to satisfy several "Performance/ Interaction Expectations."

Service technicians who fail to provide high-quality service work, fail to continually improve themselves while also helping the team continually improve, fail to treat customers well, and fail to prevent/resolve conflict within the team will not fit in well. However, those who do these things and do them well will fit in. Consequently, the team charter is a powerful tool for determining *how* to fit in at XYZ, Inc.

Job Description as a Big-Picture Tool

The *how* aspects of fitting into an organization are always about meeting expectations. You fit into an organization by doing a good job of meeting its expectations. One of the best tools for helping you understand the organization's expectations of you is your job description. As an interesting aside, if the job description is written properly, it will also show *where* you fit in. Every position in an organization has (or should have) a job description that explains the duties and responsibilities of the position or, said another way, what the organization expects of you.

Figure 4-5 shows the type of job description an electronics technician working as a service technician at XYZ, Inc. would have. By reviewing this job description, you can determine both and how and where a service technician fits into XYZ, Inc.

The first paragraph explains where service technicians fit in at XYZ, Inc. and who is in their reporting structure (often referred to as the chain of command). As a technician in the company's Service department, you report directly to a service team leader, who reports to the director of the

---

XYZ, Inc.
Job Description
ELECTRONICS TECHNICIAN: SERVICE DEPARTMENT

Individuals in this position are assigned to a Service Team in the Service department of the company's Manufacturing division. The reporting structure for this position is as follows:

> *Supervisor*: Service Team Leader
> *Manager*: Director of the Service Department
> *Vice-President*: Vice-President of the Manufacturing Division

Duties and Responsibilities:

1. On-site maintenance, troubleshooting, and repair of customers' equipment.
2. Regular maintenance and upkeep of all tools and equipment provided to you by the company.
3. Mutual support of other members of the service team to which you are assigned as appropriate.
4. Establishment and maintenance of positive, friendly working relationships with customers while projecting an image of professionalism at all times.
5. Other duties and responsibilities as assigned by your service team leader.

Qualifications:

Individuals in this position must have *one* of the following credentials:
1) Associate Degree in Electronics, 2) Technical Certificate in Electronics, or
3) Well documented equivalent training in military schools or on the job.

FIGURE 4-5 A well-written job description will help show where and how you fit into your employer's organization.

Service department, who reports to the vice-president of the Manufacturing division. You might recall from the organizational chart for XYZ, Inc. that the vice-president of the Manufacturing division in turn reports to the company's president and CEO.

How you will fit in as a part of XYZ's Service department is explained under the heading "Duties and Responsibilities." Electronics technicians who do a good job of performing the duties and responsibilities listed under this heading will fit in well at XYZ, Inc. Notice that the last entry under "Duties and Responsibilities" is what is known as the "other duties clause" of the job description.

Most job descriptions have such a clause, and all job descriptions should. This clause gives supervisors the discretion to assign team members tasks that are not specifically covered in their job description but are important to the team. For example, the newest person on a team is sometimes assigned the more menial chores that must be done but that are not spelled out in the job description (running errands for the team, cleaning up at the end of the work day, making coffee, and so on). The supervisor can invoke the "other duties clause" in the job description for more important tasks also, such as asking a team member to represent the team at a company meeting or to make a presentation to higher management.

Performance Appraisal as a Big-Picture Tool

Another valuable tool for determining how you fit into an organization is the form used to periodically evaluate your performance. This form is called the performance-appraisal instrument. It contains the various performance criteria your supervisor applies when evaluating how well you are doing your job. Figure 4-6 is an excerpt from a performance appraisal instrument used by the hypothetical company, XYZ, Inc.

This excerpt contains just four criteria provided as examples. A full instrument would contain criteria relating to all of the organization's expectations concerning your performance on the job. If you were an electronics technician in the Service department at XYZ, Inc., the company's performance appraisal form would show the criteria you would have to satisfy to fit in as a positive, contributing member of the team.

For example, look at the criteria in the excerpt in Figure 4-6. To fit in as a contributing member of your service team, you must do a good job of troubleshooting the equipment of customers, making adaptations to equipment so that it is tailored to customer needs, being a good team player, and maintaining positive, friendly customer relations. Technicians who do these things well will be valued by their supervisor, teammates, and the company.

Excerpt from a
PERFORMANCE APPRAISAL INSTRUMENT
XYZ, Inc.

Rate the employee in question on each criterion according to the following scale:

5 = Always True
4 = Usually True
3 = Sometimes True
2 = Somewhat False
1 = Usually False
0 = Always False

_____ Competently performs troubleshooting tasks.
_____ Competently adapts technology and equipment to meet the specific needs of customers.
_____ Is a good team player.
_____ Maintains positive, friendly customer relations.

FIGURE 4-6 Performance appraisal instruments show what is important enough to your employer to be evaluated.

Review Questions

1. Explain the concept of an organization's "big picture" as it relates to you as an electronics technician.
2. Would you use your employer's organizational chart to determine how or where you fit in? Explain.
3. What is a team charter, and how can you use one to determine how and where you fit into an organization?
4. What does your job description tell you about how and where you fit into an organization?
5. Explain how you would use a performance appraisal instrument to determine how to fit into an organization.

Discussion Questions

1. Discuss the following question with your classmates: "Why do you think it is important for electronics technicians to understand how and where they fit into an organization?"

2. Discuss the following question with your classmates: "Why is it important for employers to make sure that all personnel know how and where they fit into the organization?"

3. Defend or refute the following statement: "When a new problem or an unexpected crisis occurs on the job, I am better able to make good decisions about how to handle the situation if I understand my organization's big picture as well as how and where I fit into it."

Application Assignments

1. Locate an organizational chart for a company that hires electronics technicians, and determine which functional units they might fit into.

2. Go online and locate a job description for an electronics technician. Analyze the job description to determine how an electronics technician would fit into the organization in question.

3. Locate a performance appraisal instrument for an organization that hires electronics technicians. Examine the instrument to determine how a technician would fit into the organization in question.

4. Identify an individual who works as an electronics technician. Ask this technician to explain where his or her department fits into the larger organization in question and where he or she fits into the department.

CHAPTER 5

APPLY SELF-DISCIPLINE AND GOOD TIME MANAGEMENT

LEARNING OBJECTIVES

Upon completion of this chapter, you should be able to do the following:

- *Define the term* self-discipline *as it applies to electronics technicians.*
- *Explain the importance of self-discipline to electronics technicians who are trying to build a successful career.*
- *Explain how self-discipline can advance your career as an electronics technician.*
- *Determine by self-assessment your personal self-discipline quotient.*
- *Demonstrate how to make effective and efficient use of your time.*

Jenny has a good job with an electronics-manufacturing firm, and she is a talented technician. As the company grows, Jenny hopes to grow with it, but she has a problem. Jenny is a poor manager of her time. As a result, she often wastes her own time as well as that of others. Jenny knows that her lack of self-discipline and poor time management might hurt her career, and she wants to do better. Unfortunately, she just does not seem to be able to break old habits. Jenny needs to develop an important new skill. That skill is self-discipline.

How many people do you know who decide to go on a diet but don't stick with it very long? How many people do you know who start an exercise program but don't workout consistently? How many people do you know who want to stop smoking but don't do it? What is missing when people know they should do something and really want to do it, but don't follow through with their commitments is self-discipline.

People who fail to follow through on their commitments often rationalize their failure by saying, "I just don't have any self-discipline." This is not a good excuse. The truth is that people who use this excuse do have self-discipline, they just don't apply it. Self-discipline is not a gift you receive at birth or a genetic trait. Self-discipline is a skill that can be developed—just like any other skill. Think of all the skills you have now in the field of electronics that you did not have when you first enrolled in your program. You had to develop those skills. The same is true for self-discipline. Electronics technicians who want to build successful careers must develop self-discipline because it takes self-discipline to do what is necessary to succeed.

What Is Self-Discipline?

Self-discipline is a skill that allows you to consciously take control of your choices, decisions, actions, and behavior. The two key concepts in this definition are represented by the words "skill" and "consciously." The concept of "skill" is important because self-discipline is a skill that can be learned and developed over time through consistent practice. The concept of "consciously" is important because it shows that self-discipline is a choice rather than an ability people are born with.

People who develop and apply self-discipline in their lives are just as prone to human weakness as anyone else. The difference between those who develop and apply self-discipline and those who don't is a conscious choice. People who do not choose to exercise self-discipline often think that making the right choice in a given situation is easier for other people. It is not. Self-disciplined people would like to take the easy, comfortable, or expedient way just as much as anyone else. The difference is that they consciously choose to do the right thing.

Self-disciplined people are self-disciplined because they choose to be. Those who are not self-disciplined are not because they choose not to be. Exercising self-discipline often means doing what you know is right or what you know you should when you would really like to do something else—something that is easier, more convenient, or more fun. In other words, self-discipline requires *self-denial*, which is what makes it so hard. We naturally want what we want and when we want it. Very few people are willing to deny themselves the comfort, expedience, or fun of the easy option in order to do what is necessary to succeed. This is another reason many people do not succeed—they lack the willingness to deny themselves comfort, ease, or expedience when necessary. This is because they fail to apply self-discipline—a conscious choice on their part.

Importance of Self-Discipline

Often the biggest difference between success and failure is self-discipline. In every organization, there are talented people who never succeed because they lack self-discipline. This is why self-discipline is so important to those who want to build a successful career in electronics. Self-discipline shows up in a number of beneficial ways in the workplace. Consider the following ways in which self-discipline will benefit you at work and in your career (see Figure 5-1).

Better Time Management

To succeed as an electronics technician, you will have to make efficient use of your time. The workplace is often hectic and almost always busy. Often, it will seem as if you just don't have time to do everything that needs to be done. Consequently, effective time management is important, and effective time management requires self-discipline. Self-disciplined technicians make effective and efficient use of their time. This, in turn, improves their individual performance and the performance

---

Benefits of Self-Discipline

- Better time management
- Good stewardship of resources
- Effective execution of plans

FIGURE 5-1 Self-discipline can help your career advancement.

of their team. Whenever you improve your performance and that of your team, you also improve your career advancement potential.

Stewardship

The most successful electronics technicians are good stewards of the resources entrusted to them (for example, tools, equipment, supplies, and so on). This means they take good care of these resources and use them wisely which, of course, takes time. Because self-disciplined electronics technicians manage their time well, they have the time necessary to be good stewards. This is important because being a good steward will increase your career advancement potential.

Execution

Self-disciplined electronics technicians are more efficient and effective in executing the parts of their employer's plans they are responsible for. Even the best plans in the world are just dreams until they are executed. Effective execution of plans requires self-disciplined team members who can stay on time and on task. This is important because in a competitive environment, the difference between winning and losing is often determined by how effectively an organization executes its strategic and operational plans. Electronics technicians who earn a reputation for doing their part to execute their employer's plans go farther in their careers, and they do so faster.

How Self-Discipline Can Help Advance Your Career

The value of self-discipline can be seen in the following scenario involving two electronics technicians. Mary and Jane are electronics technicians for the same company. Although their resumes look almost exactly alike, they perform at noticeably different levels on the job. Mary typically has the highest performance rating in her team. Jane, on the other hand, typically has the lowest. The key difference between these two electronics technicians is that Mary is a good, self-disciplined time manager.

Mary always arrives at work on time, plans her activities before beginning a new project, sets a schedule for herself, and sticks to the schedule. This approach to her work allows Mary to make effective and efficient use of her time. It also leaves her with time in her schedule to be a good steward—properly maintaining her tools and equipment and carefully controlling her inventory of supplies.

Jane, on the other hand, is undisciplined and a poor time manager. She often oversleeps and, as a result, arrives at work either late or just barely on time. Because she is often running behind schedule in arriving at work, Jane typically jumps right into her work projects without any planning or scheduling. In addition, she spends a lot of time during the workday socializing with friends, talking on the telephone, and playing games on the Internet. Consequently, Jane is always behind in her work, which leaves her no time to be a good steward. Because she does not have time to take care of her tools, equipment, or inventory of supplies, Jane is often forced to deal with malfunctions and breakdowns. Also, it seems that she constantly runs short of critical supplies at the worst possible times.

Mary and Jane are equal in terms of their education and electronics skills, but Mary is much more self-disciplined and a better time manager. As a result, she is enjoying a much more successful career than Jane. In fact, within just two years of working together, Mary became Jane's team leader and supervisor.

Self-Discipline Assessment

Are you a self-disciplined person? You need to be. Find out if you have developed self-discipline by completing the following self-assessment. Ask yourself each of the following questions. The desired answer to each question is "Yes."

1. Do you consistently arrive at classes on time or early?
2. Do you consistently arrive on time or early for appointments with teachers, counselors, and other students?
3. Do you consistently submit assignments on time or early?
4. Do you consistently keep up with other school-related responsibilities?
5. Are you consistently on time for work (if applicable)?
6. Do you consistently and promptly return telephone calls?
7. Do you consistently and promptly return e-mail messages?
8. Do you consistently arrive on time for dates or other social events?
9. Do you consistently arrive on time for family events?
10. Do you consistently follow through on promises you make?

If you can honestly answer "Yes" to all 10 of these questions, you have a *self-discipline quotient* of 100 percent. Any question you answer "No" to indicates a need to improve your self-discipline. The most important word in each question is "consistently." In order to answer "Yes," you must consistently do what is asked.

Good Time Management

The most successful electronics technicians use their time wisely—which means they use it efficiently and effectively. This is important because technicians who poorly manage their time often find themselves falling short on critical tasks as a result. Poor time management can cause problems on the job such as wasted time, stress, missed appointments, missed deadlines, insufficient attention to detail (which leads to mistakes), ineffective execution of plans, and poor stewardship of resources to name just a few. Consequently, it is important for you to become a good time manger.

Time-Management Challenges and Their Solutions

Here is a common scenario in today's workplace. An electronics technician comes to work with a list of tasks that should be accomplished today. Unfortunately, he hardly even gets started on his work before his time and efforts must be diverted to unplanned, unscheduled tasks that are not on his list. The reasons that unplanned, unscheduled activities rob technicians of so much time on the job vary, but most of them are predictable.

Common time-management challenges for electronics technicians include emergencies, telephone calls, poor planning, taking on too much, disorganization, and inefficient use of technology (see Figure 5-2). The most successful electronics technicians have learned how to overcome these challenges by applying good time-management strategies. Time-management strategies that will help you manage your time better follow in this section.

Checklist of:
TIME MANAGEMENT CHALLENGES

- Emergencies

- Telephone calls

- Poor planning or no planning

- Taking on too much

- Personal disorganization

- Inefficient use of technology

FIGURE 5-2 These factors will inhibit effective time management.

Emergencies

Emergencies are going to happen. A rush job from an important customer is going to come in and take precedence over everything else you had planned. Equipment you need to use is going to break down. A teammate who is working on an important project is going to call in sick or leave for another job. These things are going to happen, and when they do, plans and schedules typically go out the window. While it is true that the better you plan, the fewer emergencies you will face, it is also true that even with the best of plans, emergencies are going to happen.

Events over which you have no control can create emergencies that must be dealt with. For example, a member of your team might be the victim of an automobile accident. There was nothing you could do to predict or prevent this unfortunate occurrence. It happened and, as a result, you and your teammates are going to have to double up and do her work as well as yours—and the deadline has not changed. The following strategies will help you deal as effectively as possible with the emergencies that inevitably occur at work:

- *Leave some unscheduled time in your day.* When developing your work schedule for the day, rather than schedule every minute, leave some unscheduled time. Then, when emergencies arise, you will have at least some time to deal with them and still get your scheduled work done. If emergencies do not arise in a given day, you can always put the unscheduled time to good use by completing scheduled work, getting ahead of schedule on projects, or performing routine maintenance tasks on your tools, equipment, and work station.

- *Avoid getting bogged down in gossip and gripe sessions.* No matter where you work, there will be people who spend an inordinate amount of time during the workday gossiping about others or griping about certain aspects of the job. You should studiously avoid gossip and gripe sessions for a number of reasons; the pertinent one in the current context being that they typically take up a lot of time but do no good for anyone.

Telephone Calls

Unless you manage its use wisely, the telephone can rob you of valuable time; especially cell phones. Because cell phones allow friends, customers, suppliers, teammates, and anyone else to call you any time no matter where you are or what you are doing, they tend to waste a lot of your time. To keep your telephone from becoming a time robber, use it only for business during the workday unless there is a family emergency. Also, when using the telephone at work, avoid idle chatter; get to the point, and be brief.

Poor Planning or No Planning

A good time-management rule to remember is this: *A little planning can save a lot of time.* The best way to overcome the planning problem is to end each day by planning for the next. Never come to work wondering what you are supposed to that day. Before going home that day, take stock of what needs to be done the next day. Make yourself a "to do" list, and put the items on it in order of priority—the most important task first, the second most important next, and so on. In this way, you will know exactly what needs to be done when you walk in the door the next day, and, if emergencies occur, you will know which tasks on your list are most important and which can be put off.

Taking On Too Much

One of the reasons successful electronics technicians are successful is that they take the initiative and seek responsibility when they see work that needs to be done. This is good and I recommend this approach to all technicians who want to build a winning career. However, balance and common sense are important here. If you are already overloaded and having difficulty meeting a deadline, it is not a good time to seek additional responsibility. Use some common sense. If you are making good progress with your assigned work, and you see something in the team that needs to be done, take the initiative and get it done. However, if taking on new work is just going to cause you to miss the deadline for your current assignments, hold off and stick to completing what is currently on your plate.

Personal Disorganization

You can waste a lot of time rummaging around looking for tools, parts, or supplies that aren't where they are supposed to be. I know an electronics technician who has excellent technical skills but has not advanced as would be expected in his career because he is so disorganized. For example, he never puts the tools or devices he uses back where they belong. Rather, he just lays them down wherever he happens to be when he is done with them. In addition, he constantly misplaces paperwork relating to his assigned projects.

The source of this individual's disorganization is a lack of self-discipline, as it is with most disorganized people. He is disorganized because he does not discipline himself to take the time necessary to put things where they are supposed to be, file paperwork, inventory supplies, or organize his work station at the end of each day.

There is a running joke about this individual among his teammates concerning his car keys. This individual wastes an inordinate amount of time each day searching for his "lost" car keys simply because he never puts them in the same place when he arrives at work. This is just another example of his failure to discipline himself. Personal disorganization would be a bad enough problem if it wasted only your time, but that is not how things happen in a team. When a technician is disorganized, he typically wastes not only his time but that of teammates, customers, and suppliers.

If you are disorganized, make a point of disciplining yourself to 1) assign a place for your tools and equipment and never go home at the end of the day until you have returned them to their place; 2) straighten up your work station and get it organized for tomorrow before leaving today; 3) establish a simple and convenient file system for your paperwork and use it; and 4) at least four times per year, get a large trashcan and clean out old files, drawers, and all other materials at your workstation that are no longer needed or valid.

Inefficient Use of Technology

Even well-trained electronics technicians are sometimes guilty of inefficient use of technology. Technologies such as the computer, cell phones, and many of the devices you will use for troubleshooting electronic equipment are designed with features that are supposed to save time for users. However, these features will save time only if you know how to use them and use them well. If you don't, technology is just one more thing that robs you of time.

Make a list of all the technologies you use in your job (for example, computer, cell phone, troubleshooting devices, and so on). Now ask yourself if you can also list all of the various time-saving features of each device. Take the time to learn all of the features that are available to you as well as how to use them. They won't save you any time unless you know what they are and how to use them.

Review Questions

1. Define the term *self-discipline* as it applies to electronics technicians.
2. Explain why self-discipline is so important to electronics technicians who are trying to build a successful career.
3. How can you apply self-discipline to help advance your career in electronics?
4. Why is it important to conduct a self-assessment to identify your self-discipline quotient?
5. List five common time wasters at work and what you can do about them.

Discussion Questions

1. Have a discussion with classmates about self-discipline. Without giving names, talk about people you have known who lacked self-discipline and any problems this caused.

2. Defend or refute the following statement: "Self-discipline is something you either have or you don't. There is nothing you can do about it if you don't have it."

3. Have a discussion with classmates about time management. What are some time-management problems you have to face and how do you handle them?

Application Assignments

1. Conduct a self-assessment of your self-discipline quotient. Based on the results of this self-assessment, develop a plan for improvement.

2. Make a log for one week of time-management problems that come up in your life or that you observe happening around you. For each problem observed, make notes concerning how the problem could have been prevented or the impact of it decreased.

3. Identify someone who works in electronics. Ask this person to tell you how self-discipline and good time management affect the flow of work on a day-to-day basis (both good and bad). Ask for specific examples that you can share with classmates.

CHAPTER 6

BE AN EFFECTIVE TEAM PLAYER, TEAM BUILDER, AND TEAM LEADER

LEARNING OBJECTIVES

Upon completion of this chapter, you should be able to do the following:

- *Define the term* team *as it applies to electronics technicians.*
- *Explain the responsibilities of team players, team builders, and team leaders.*
- *Demonstrate how to apply the four-step model for team building.*
- *Explain the difference between "coaching" and "bossing" teams.*
- *Explain how to manage conflict in teams.*

Teamwork is fundamental to your employer's success in a competitive environment. Consequently, being a good team player is fundamental to your success as an electronics technician. The reason that teamwork is so important to employers can be summarized in one word: *competition*. To succeed in today's hyper-competitive global business climate, employers need electronics technicians who can make their team perform better by being good team players. The better the performance of its teams, the better the performance of the employer, and the more competitive it will be. Therefore, you will want to begin your career in electronics as a good team player. Then as you advance, you will want to become a good team builder and team leader.

What Is a Team?

A *team* is a group of people with a common purpose. Having a common purpose is one of the keys to effective teamwork. The truth in this point can be seen in the performance of sports teams. For example, a basketball team in which one player hogs the ball and pursues his own personal goals (MVP, publicity, or something else) will rarely win against a team of players, all of whom work together toward the common goal of winning the game. Before you can be a team builder or team leader, you must first be a team player. Team players unselfishly help their teams win—in sports and on the job. People who are able to help teams win are more likely to become winners themselves. This chapter will help prepare you to become a team player, team builder, and team leader.

Responsibilities of Team Players, Builders, and Leaders

To be a good team player and develop into a good team builder and team leader, you need to know what is involved—what the responsibilities are in each of these areas. Although they overlap, the responsibilities relating to being a team player, team builder, and team leader are different.

Responsibilities of Team Players

A good team player is willing to put the goals of the team ahead of any personal agenda. Good team players think: "Team first, me second." What follows are the responsibilities of good team players:

- Understand the team's charter (mission, goals, and guiding principles).
- Know what must be done to accomplish the team's mission and goals.

- Be open and honest with teammates and the team leader.
- Work to build trust between and among team members.
- Cooperate with teammates in getting the work of the team completed.
- Take responsibility for the team's performance.
- Maintain a positive attitude toward the team and its work.
- Take the initiative to enhance the team's performance.
- Be resourceful in finding ways to get the job done when the team runs into difficulties.
- Set an example of persevering when the team's work becomes difficult.
- When you disagree with teammates, don't be disagreeable.
- Know how team success, progress, and performance are measured (accountability).
- Understand how decisions are made in the team and, after they are made, do your part to carry them out even if you disagree with them (unless there is a question of ethics or legality).

Responsibilities of Team Builders

Team builders are team players who work to continually improve the performance of the team. The responsibilities of good team builders are as follows:

- Make sure that other team members understand the team's charter.
- Constantly assess the team's performance to identify weaknesses that should be corrected.
- Based on assessments, plan activities to strengthen the team's abilities.
- Execute the plans you develop for strengthening the team's abilities.
- Monitor and evaluate to determine how effective your team-building activities have been.
- Help resolve conflict in the team before it becomes an issue.

Responsibilities of Team Leaders

Becoming a team leader is an important step up the career ladder. Consequently, when you get an opportunity to lead a team of electronics technicians, it is important to get it right. What follows are the principal responsibilities of team leaders:

- Consistently set a positive example of the attitudes and behaviors you expect of team members.
- Serve as the team's coach.
- Connect the team with the rest of the organization.
- Ensure that all team members understand the team's charter and their roles in carrying it out.

- Inspire and motivate team members to achieve consistent peak performance.
- Work with higher management to ensure that team members have the training, resources, and support they need to do the job that is expected of them.
- Monitor, evaluate, and continually improve team performance.

Creating the Team's Charter

A team's charter consists of a mission statement and a list of ground rules (see Figure 6-1). The team's mission statement explains what it does—the team's purpose. The mission statement explains the team's reason for being. A mission statement is written in terms that are broad enough to encompass everything the team will be expected to do but specific enough to allow progress to be measured. The sample mission statement in Figure 6-1 meets both of these criteria.

This statement is broad enough to encompass all of the activities that go into assembling power supplies. On the other hand, it is specific enough that the performance of the team in question can be measured

---

Team Charter
ELECTRONIC ASSEMBLY TEAM

Mission Statement
The mission of the Electronic Assembly Team is to assemble high-quality power supplies for avionics applications.

Team Ground Rules

1. Our aim is consistent peak performance and continual improvement. We expect the best of all team members every day.
2. All team members are expected to work together in a mutually supportive way.
3. All team members are expected to put the goals of the team ahead of personal agendas.
4. All team members are expected to resolve disagreements among themselves without being disagreeable and based on what is best for the team's performance.
5. All team members are expected to maintain a positive attitude toward their work, the team, each other, and the organization.

FIGURE 6-1 Team charters provide direction for team members.

(for example, how many power supplies are produced daily, weekly, or monthly, and to what level of quality?). The sample mission statement is written in broad terms but is specific enough that team members know they are expected to focus simultaneously on both quality and productivity. It also meets one other important criterion—simplicity. Any employee could understand this mission statement. It is brief and to the point, but comprehensive.

Team leaders should keep these criteria in mind when developing mission statements for their team charters: broadness, specificity, and simplicity. A good mission statement is a tool for communicating the team's purpose, both within the team and throughout the organization.

The ground rules in the team charter explain how team members are supposed to approach their work as well as how they are supposed to interact with each other. Members of the electronic assembly team who use the team charter in Figure 6-1 will know that they are expected to give their best effort every day on every project (consistent peak performance), get better at their jobs all the time (continual improvement), work well with teammates in a mutually supportive manner, put personal agendas aside in favor of achieving team goals, resolve conflicts according to what is best for the team, and maintain a positive can-do attitude.

Four-Step Approach to Team Building

When you become a team leader, building your team will be an on-going, everyday process that never stops. Even when your team begins to function well, it can still get better. Effective team building is a four-step process:

1. Assess
2. Plan
3. Execute
4. Evaluate

To be a little more specific, the team building process proceeds along the following lines 1) assess the team's developmental needs (its strengths and weaknesses), 2) plan team-building activities based on the needs identified, 3) execute the planned team building activities, and 4) evaluate results. These steps are explained in the following sections.

Assessing Team Needs

If you were the new coach of a basketball team about which you knew very little, what is the first thing you would want to do? Most coaches in such situations would begin by assessing the abilities of their team members. What are our weaknesses? What are our strengths? With these questions answered, the coach would know how best to proceed with team-building activities.

This same approach can be used in the workplace. A mistake often made by team leaders is beginning team-building activities without first assessing the team's strengths and weaknesses. Resources are limited in organizations. Consequently, it is important to use them as efficiently and effectively as possible. Team leaders who begin team-building activities without first assessing strengths and weaknesses run the risk of wasting resources by providing training and mentoring to strengthen characteristics that are already strong. Team leaders who do this use up the limited resources they need to improve weaknesses. Assessing team needs before planning team-building activities will give you an advantage over other team leaders who skip this important step.

For workplace teams to be successful, they should have at least the following characteristics:

- Clear direction that is understood by all members (team charter)
- "Team players" on the team (team first—me second)
- Fully understood and accepted accountability measures (evaluation of performance)

Figure 6-2 is an assessment tool that can be used for identifying strengths and weaknesses in teams.

Planning Team-Building Activities

Team-building activities should be planned based on the results of your assessment of needs. For example, say the team appears to be having problems because it lacks direction. Clearly, part of the process of building this team must be either developing a mission statement or explaining the existing statement. If team members don't seem to be as interested in their performance as they should be, accountability measures might be weak. Regardless of what your assessment reveals about the team, use that information to plan team-building activities that will enhance the team's performance.

Executing Team-Building Activities

When you become a team leader, remember that team-building activities should be implemented on a just-in-time basis. This means it is best to provide team-building activities as the need for them is identified. Do not wait. Like any kind of training, teamwork training is best provided when the need is apparent and immediate. Consequently, the best time to provide teamwork training is after an assessment has been conducted. In this way, the need is fresh on the minds of team members. Team building is an ongoing process. The idea is to make a team better and better as time goes by.

Assessment Instrument
TEAM STRENGTHS AND WEAKNESSES

Instructions

To the left of each item is a blank for recording your perceptions. For each item, record your perception of how well it represents reality in your team using the following code:

> 6 = *Completely true*
> 4 = *Somewhat true*
> 2 = *Somewhat false*
> 0 = *Completely false*
> X = *Not applicable*

Understanding of the Team's Mission and Ground Rules

_____ 1.   The team has a clearly stated mission.
_____ 2.   All team members understand the team's mission.
_____ 3.   The team has clearly stated ground rules.
_____ 4.   All team members understand the ground rules.
_____ 5.   All team members consistently comply with the ground rules.

Understanding of Team Accountability Measures

_____ 6.   All team members know what is expected of them in terms of performance.
_____ 7.   All team members know how their performance and that of the team is measured.

Possession of Positive Team Player Characteristics

_____ 8.   All team members trust each other.
_____ 9.   All team members can depend on each other.
_____ 10.   All team members treat each other with mutual respect.
_____ 11.   All team members are punctual.
_____ 12.   All team members take responsibility for their performance and that of the team.
_____ 13.   All team members take the initiative to continually improve performance.
_____ 14.   All team members support decisions once they are made.
_____ 15.   All team members resolve conflict within the team without becoming disagreeable.

FIGURE 6-2 Always assess before planning for improvement.

Evaluating Team-Building Activities

If team-building activities have been effective, weak areas pointed out by the needs assessment should have been strengthened. An easy way to evaluate the effectiveness of team-building activities is to wait a sufficient amount of time to allow the team's performance to improve and then conduct the needs assessment again.

If this subsequent assessment shows that sufficient progress has been made, nothing more is required. If not, additional team-building activities are needed. If a given team-building activity appears to have been ineffective, get the team together and discuss it. Use the feedback from team members to identify weaknesses and problems, and use the information to ensure that team-building activities become more effective. Involving the team in evaluating team-building activities is itself an excellent team-building activity.

Be a "Coach" Not a "Boss"

Teams are not bossed—they are coached. Team leaders need to understand the difference between bossing and coaching. *Bossing*, in the traditional sense, involves giving orders and evaluating performance. Bosses approach the job from an "I'm in charge—do what I say" perspective. Bosses are said to want team members to work from the "neck down." In other words, they want them to just do what they are told without thinking.

Coaches, although clearly in charge, approach the job as facilitators of team performance, team development, and team improvement. They lead the team in such a way that it achieves peak performance on a consistent basis. They want team members to think as well as work. You can be an effective coach of work teams by doing the following:

- Giving your team a clearly defined charter (refer to Figure 6-1)
- Making team development and team building a continual activity
- Mentoring team members
- Promoting mutual respect between you and team members and among team members
- Working to make human diversity within a team a plus

Clearly Defined Charter

You can imagine a sports coach calling her team together and saying, "This year we have one overriding purpose—to win the championship." In one simple statement, this coach has clearly and succinctly defined the team's mission (part of the team charter). All team members now know that everything they do this season should be directed toward winning the championship. Coaches of work teams should be just as specific in explaining the team's mission to their team members.

Team Development/Team Building

Work teams should be similar to athletic teams when it comes to team development and building. Regardless of the sport, athletic teams practice constantly. During practice, coaches work on developing the skills of individual team members and the team as a whole. Team development and team-building activities are a constant, and they should go on forever. A team should never stop getting better. Coaches of work teams should follow the lead of their athletic counterparts. Developing the skills of individual team members and building the team as a whole should be a normal part of the job—a part that takes place regularly and never stops.

Mentoring

Good coaches are mentors. This means they establish helping, caring, nurturing relationships with team members. Developing the capabilities of team members, improving the contribution individuals make to the team, and helping team members advance their careers are all mentoring activities.

Mutual Respect

It is important for team members to respect their coach, for the coach to respect his team members, and for team members to respect each other. There must be mutual respect for a team to function at peak levels on a consistent basis.

Human Diversity

Human diversity can be a plus in teams. Leading organizations in the United States have learned that most of the growth in the workplace for the foreseeable future will be from women, minorities, and immigrants. These people will bring new ideas and varying perspectives, precisely what an organization needs to stay on the razor's edge of competitiveness. However, in spite of steps already taken toward making the American workplace both diverse and harmonious, wise coaches understand that people—consciously and unconsciously—sometimes erect barriers between themselves and people who are different from them. This tendency can quickly undermine the trust and cohesiveness on which teamwork is built. Wise coaches work hard to keep this from happening on their teams. This is such an important topic that an entire chapter is devoted to it later in this book (Chapter 10).

Resolving Conflict in Teams

Conflict will occur in even the best of teams. Even when all team members agree on a goal, they can still disagree on how best to accomplish it. Few things will further your career more than being able to resolve conflict in a positive way. An entire chapter is devoted to this important topic later in this book (Chapter 14).

Review Questions

1. In your own words, define the term *team*.
2. Explain how to choose team members when forming a work team.
3. List and explain the main responsibilities of team leaders.
4. Write an example of a mission statement for a hypothetical team's charter.
5. Explain the four-step approach to team building.
6. What is the difference between a *boss* and a *coach* in the language of teamwork?

Discussion Questions

1. Discuss the benefits of having a team charter if you were a new member of a team of electronics technicians.
2. Would you rather work on a team that is coached or a team that is bossed? Discuss your reasons.

Application Assignments

1. Think of a team you have served on. It can be a work team, sports team, or a project team in school. Use the assessment instrument in Figure 6-2 to identify the strengths and weaknesses of that team.
2. Develop a team charter for a team of electronics technicians in a hypothetical company.
3. Identify someone who works in electronics, and ask this person to tell you how teamwork comes into play in the workplace (both good teamwork and the lack of teamwork). Ask for specific examples that you can share with your classmates.

CHAPTER 7

BE AN EFFECTIVE COMMUNICATOR

LEARNING OBJECTIVES

Upon completion of this chapter, you should be able to do the following:

- *Explain the importance of effective communication to a successful career.*
- *Define communication as a concept.*
- *Distinguish between communication and effective communication.*
- *Describe the communication process.*
- *List and explain the most common inhibitors of effective communication.*
- *Describe how you can improve communication by building trust.*
- *Explain the importance of listening in communication.*
- *Describe the strategies for being a good listener.*
- *Explain the importance of nonverbal communication as part of the broader concept of communication.*
- *Explain the strategies for improving your verbal communication.*

Communication is not a new topic for students of electronics. After all, much of what you study has to do with various communications technologies. For example, in school—depending on the specifics of your program—you might have studied or be studying such subjects as signal amplification technology, wireless technology, telecommunications technologies, and antenna technologies. All of these technologies are related to communication. In one way or another, they are part of the various mediums used to convey information electronically. As you will see later in this chapter, the communication medium is an important part of the communication process. But, the most important component in the process is you—the person who sends and receives the information in question.

Few things will contribute more to the success of your career as an electronics technician than the ability to communicate effectively. Communication skills will help you accurately and effectively convey information to other people on and off the job. These same skills will also help you accurately perceive information that is being conveyed to you. Effective communication is critical to your success and to the success of your employer because communication is the "oil" in the machine of human interaction. Of all the skills needed for success in the workplace, communication is among the most important. It is an essential element in the job of the electronics technician. Consequently, this chapter explains how to develop the communication skills you will need to build a winning career as an electronics technician.

Communication: An Important Skill

Communication may be the most imperfect of all human processes. It is also one of the most important. This is because the quality of communication is affected by so many different factors (for example, speaking ability; hearing ability; language barriers; differing perceptions or meanings based on age, gender, race, nationality, or culture; attitudes; nonverbal cues; level of trust between sender and receiver; and many others). Because of this, effective communication can be difficult in even the best of circumstances. If you have ever had difficulty making someone understand what you are trying to say or trying to understand what someone else was trying to say, you know what I mean. In spite of the difficulties, learning to be a good communicator is an essential step on your path to a successful career. Fortunately, communication skills are like other work-related skills in that they can be learned. With sufficient study and practice, most people—regardless of their innate capabilities—can learn to communicate well.

An electronics technician I'll call John learned about the value of good communication his first day on the job. During his orientation, the company's human resources director explained about wages, benefits, sick leave, vacation time, the dress code, and numerous other issues of importance to new employees. Anxious to get started and bored with all of the paperwork required, John tuned out. As a result, he was not listening when the human resources director explained about the dress code.

Because electronics technicians in this company have frequent contact with customers, management insists on a strict dress code to project a professional image. All technicians are required to wear what amounts to a khaki uniform with a shirt bearing the company logo over one pocket and the employee's name over the other. The company maintains an account with a local clothing store that provides the uniforms for new electronics technicians. All new technicians receive five new uniforms, which the company pays for, but they must go to the store in question to pick up the uniforms. The clothing store embroiders the company's and the employee's name on the shirts and will alter trousers and shirts to ensure a good fit. The company likes to maintain a positive image with its customers and believes that personal appearance is important.

During the part of the orientation that dealt with the dress code, John had tuned out—he simply was not listening. However, after the meeting, he vaguely remembered hearing something about wearing khakis, so he wore khakis to work the next day. Unfortunately, his shirt was missing the company logo and his name. John quickly realized that he was inappropriately dressed. In fact, compared with his fellow technicians he appeared to be "out of uniform." His new supervisor was not pleased. The other new employees were able to join their teams and begin learning their jobs. John, on the other hand, had to—in the words of his supervisor—"waste half a day going to the clothing store and picking up his uniforms." John committed one of the worst mistakes a new employee can make. He made a bad first impression—all because he failed to listen.

Communication Defined

People sometimes confuse *telling* with *communicating*. Then, when a problem develops, they are likely to say: "But I told him what to do." In addition, people will occasionally confuse *hearing* with *listening*. They are likely to say, "That isn't what I said. I know you heard me because you were standing right next to me!"

In both cases, the individual in question has confused telling and hearing with communicating. What you say is not necessarily what the other person hears, and what the other person hears is not necessarily what you intended to say. In both examples, the missing ingredient is

---

Communication

Information + Received + Fully Understood = Communication

FIGURE 7-1 Communication is an important success skill.

comprehension. What was said was heard, but it was not understood. Communication may involve telling, but it is not *just* telling. It may involve hearing, but it is not *just* hearing. With this in mind, I define communication as follows (see Figure 7-1):

Communication is the transfer of information that is received and fully understood from one source to another.

A message can be sent by one person and received by another, but until the message is understood, there is no communication. This applies to spoken, written, and nonverbal messages. In order for there to be communication, there must be comprehension (understanding).

Effective Communication

Communication in the workplace is essential, but just communicating is not enough. The most successful technicians take the time and put forth the effort to ensure *effective communication*. As was stated in the previous section, when the information conveyed is received and understood, communication has occurred. However, understanding alone does not necessarily lead to effective communication. Effective communication occurs when the information received is not just understood but also accepted.

For example, a team leader asked the technicians on her team to use a different component on the next lot of printed circuit boards they assembled. All of the team members verified that they received and understood the message. Unfortunately, the team leader conveyed the message as a directive rather than giving her technicians the reason for the change. Consequently, two technicians decided to use up the old components in their parts bins before beginning to use the new components. When the printed circuit boards containing the old components failed an intermediate quality control test, the team leader was puzzled. She had specifically told every member of her team to change to the new component. After tracing the defective circuit boards back to the two technicians in question, the team leader asked for an explanation. She was truly shocked. After all, these technicians were two of her best team members. She knew they would not knowingly install defective components.

As it turned out, they understood the directive to change to a new component, but because they did not understand why, they thought they should use up the old components in their parts bins first and then change to the new component. They were trying to economize and cut down on waste. What the team leader had failed to tell these two technicians and the rest of her team is that she had just learned that the old component was defective—thus the directive to make the change. Because these two high-performing technicians were operating on incomplete information, they did not realize the significance of their well-intended decision to use up their supply of the old component before changing to the new. There was communication in this case but not effective communication.

Effective communication is a higher level of communication because it implies not just understanding but acceptance. The acceptance aspect of effective communication can require sharing the rationale behind the message being communicated and sometimes even persuasion. On occasion, to ensure effective communication, it might be necessary to explain the "why" behind the message. Had the two technicians in this example understood why they needed to change to a new component, they would have done so immediately. They were not defying their team leader. On the contrary, these two high-performing technicians were trying to save money for their company. Remember this when attempting to communicate with others. Make sure that they receive, understand, and accept the message you are conveying.

The Communication Process

Communication is a process consisting of several components: *sender, receiver, medium,* and the *message* itself (see Figure 7-2). The sender is the originator or source of the message. When you talk to someone, you are a sender. When you e-mail someone, you are a sender. When you give someone a look that says, "Do not bother me right now," you are a sender. The receiver is the person or group for whom the message is intended. When you listen to someone, check your e-mail messages, or read a memorandum, you are a receiver. The message is the information that is conveyed and is to be understood, accepted, and acted on. The medium is the method used to convey the message.

There are three basic categories of communication mediums: *verbal, nonverbal,* and *written.* All three mediums can be aided or enhanced by

Components of the

COMMUNICATION PROCESS

- Sender
- Medium
- Message
- Receiver

FIGURE 7-2 The communication process has four basic components.

technology. Verbal communication includes face-to-face conversation, telephone conversation, speeches, voice recordings, public announcements, press conferences, and other means of conveying the spoken word. Nonverbal communication includes gestures, facial expressions, voice tone, body poses, and proximity. Nonverbal communication is a powerful part of the concept. This is why it can be more difficult to communicate by telephone than in person—you lose all of your nonverbal cues except voice tone when talking on the telephone. Written communication includes letters, e-mail, memoranda, billboards, bulletin boards, manuals, books, and any other means of conveying the written word.

Technological developments continue to enhance our ability to convey information. If you have trouble visualizing communication before e-mail and cell phones, imagine living in an era before television, telephones, or even radio. Communications-related technologies such as satellites and the Internet, coupled with advances in transportation technology, are what enabled the advent of global business. Irrespective of technologies, the better you become at using all of the various communication mediums (verbal, nonverbal, and written), the more effective you will be at communicating.

Factors That Inhibit Communication

Even though communication-enhancing technologies such as cell phones, e-mail, television, and the Internet are commonplace in today's society, there are still numerous inhibitors of effective communication that cannot be overcome by technology. Electronics technicians should be familiar with these inhibitors and learn how to avoid or overcome them (see Figure 7-3).

Inhibitors of Communication

- Differences in meaning

- Insufficient trust

- Information overload

- Interference

- Listening problems

- Premature judgments

- Inaccurate assumptions

- Technological glitches

FIGURE 7-3 These factors will inhibit effective communication.

Differences in Meaning

Differences in meaning are inevitable in communication because people in the workplace have different backgrounds and levels of education. People come from different cultures, races, and nationalities. As a result, the words, gestures, and facial expressions used by people to communicate their thoughts, ideas, and feelings can have altogether different meanings. For example, an electronics technician with an Asian background might have learned to behave in a nonconfrontational manner even when in reality he disagrees with your ideas or recommendations. Another example of when there can be differences in meaning between two technicians is what I call the "generational syndrome." People of different generations, although both are speaking English, can seem to be speaking two completely different languages. You might have had this experience yourself. To overcome this inhibitor, you have to invest the time necessary to get to know the people you work with so that you understand what they mean by their words, gestures, and facial expressions.

Insufficient Trust

Insufficient trust can inhibit effective communication. If receivers don't trust senders, they might concentrate so hard on reading between the lines for a "hidden agenda" that they miss the real message. Consider this example of how a lack of trust can inhibit communication. Two electronics technicians who had been at odds with each other in high school found themselves working not just for the same company but on the same team. As high school students, these two people had run against each other for student government president. The race had been close and heated. While campaigning, words were exchanged about promises that were made but allegedly not kept, and hard feelings developed. Later, after completing their electronics technology training, they found themselves working together on the same team. Their supervisor insisted that all team members participate openly in weekly brainstorming sessions to find ways to continually improve the team's performance. Unfortunately, no matter how good an idea proposed by one of the two technicians in question might be, the other one would invariably question it or even argue against it. They just did not trust each other. Consequently, communication between them is strained at best. This is why trust building among teammates is so important.

Information Overload

Because of advances in communication technology and the rapid and continual proliferation of information, people in the workplace often find themselves with more information than they can process effectively. This is known as *information overload*. You can guard against information overload by screening, organizing, summarizing, and simplifying the information you convey to others. An electronics technician I will call Hector worked for a company that manufactures aviation components for military aircraft. Hector once earned a promotion and a raise because he made a point of organizing, summarizing, and simplifying the information he conveyed during team meetings. While other technicians on the team tended to give long-winded explanations of their ideas and recommendations—explanations that could be hard to follow and that invariably made meetings last too long—Hector made a point of being brief, to-the-point, and simple in his explanations. As a result, when someone was needed to make a presentation to the company's management team about a project that was currently in production, Hector was selected. The company's executive managers were so impressed with Hector's ability to convey complex information in a brief, simple, and understandable way that they promoted him to a special position. He continued to work as an electronics technician in his team but at a higher salary. Then, in addition, any time the company's marketing personnel had to make a technical presentation to a potential customer, Hector was assigned temporarily to marketing to help make the presentation.

Interference

Interference is any external distraction that prevents effective communication. This might be something as simple as background noise caused by people talking or as complex as atmospheric interference with satellite communications. For example, you might have experienced walking around outside of a building trying to find a spot with good reception on your cell phone, or you might have needed to move to a quieter location so that you could have a conversation with someone. In both cases, the problem was interference. Regardless of its nature, interference either distorts or completely blocks the message. Because of interference, you must be attentive to the environment and your surroundings when trying to communicate with employees. Michael, the leader of a team of electronics technicians, found this out one day when he tried to conduct a team meeting while the events surrounding the tragic terrorist attack of September 11, 2001 in New York City were unfolding. There were televisions and radios on throughout the company so people could follow developments. Consequently, rather than listening to Michael as he attempted to conduct his meeting, the electronics technicians in his team were focused on listening to the radio or watching the nearest television. Noticing that his team members were distracted, Michael finally decided to cancel the meeting and reschedule it for another time. This was the best way he could think of at the time to overcome the distraction that was interfering with his meeting.

Listening Problems

Listening problems are one of the most serious inhibitors of effective communication. They can result from both the sender not listening to the receiver and vise versa. To be a good communicator, you must be a good listener. Joyce is the service director for a firm that troubleshoots and repairs computers and peripheral equipment. As an electronics technician, Joyce was a good listener—a fact that helped her successfully climb the career ladder from technician to service director. When customers would bring in a computer or peripheral to be repaired, Joyce always listened attentively. She listened with her ears as well as her eyes—in other words, in addition to listening, Joyce watched carefully for nonverbal cues. She never interrupted customers. However, once they had completely explained the problem in question, Joyce always made sure that she asked questions for clarification if needed and repeated the problem back to the customer to make sure she fully understood. Her fellow technicians, on the other hand, often rushed customers, interrupted to ask questions, and even cut off their explanations in mid-sentence when they felt sure they knew what the problem was. Consequently, before long, customers began to request Joyce when they brought in work—a fact the company's owner noticed. Joyce was an excellent technician, but so were her colleagues. The difference between Joyce and her teammates had more to do with listening ability than technical ability. Joyce was eventually promoted ahead of her fellow technicians because her excellent listening skills were good for business.

Premature Judgments and Inaccurate Assumptions

Premature judgments by either the sender or the receiver not only inhibit effective communication but also lead to inaccurate assumptions. Making premature judgments is a type of listening problem because as soon as you make a quick judgment, you are prone to stop listening. Stop listening and you are prone to make inaccurate assumptions. Return to the example of Joyce in the previous paragraph. Even if a certain customer was obviously wrong when explaining the problem he was having with his computer, Joyce would still listen patiently to the explanation. Her fellow technicians, on the other hand, would not. They thought listening to an inaccurate explanation was a waste of time. Consequently, instead of listening, they would simply interrupt the customer, take the computer in for testing, and identify the problem using standard troubleshooting procedures. Joyce used the same procedures but only after listening to the customer's explanation. She understood that listening patiently was good for business and that occasionally a customer might say something that would turn out to be helpful. You cannot make premature judgments and maintain an open mind. Therefore, it is important to listen nonjudgmentally when communicating with others.

Technological Glitches

Software bugs, computer viruses, dead batteries, power outages, and software conversion problems are just a few of the technological glitches that can interfere with communication. The more dependent we become on technology for conveying messages, the more often these glitches will interfere with and inhibit effective communication. As an electronics technician or a technician in training, you understand this. In fact, troubleshooting technological glitches is likely to be part of your job as an electronics technician regardless of where you work. While you are busy identifying technological glitches that are inhibiting the communication of others, do not forget to identify any that might be causing problems with your communication. Do not be like the automobile mechanic whose car broke down because he was so busy taking care of everyone else's cars that he failed to take care of his own.

Building Trust to Improve Communication

In the previous section, the point was made that you will not be able to communicate effectively with people who don't trust you. Trust is essential to effective communication because if people don't trust you, they either will not believe what you tell them, or they will question your motive in telling them. What follows are some practical strategies you can use to build trust among the people you will need to be able to communicate with:

- When you talk to people, tell the truth. Be tactful, but tell them what you think, believe, or feel. Even when the news you have to deliver is bad, tell the truth without a lot of sugarcoating. Never leave people to wonder what you said or what you meant when you said it. Telling the truth can be difficult at times but trying to maintain a lie can be even harder.

- Don't make promises you cannot keep. Instead, promise small but deliver big. When you make a promise or commitment to someone, follow through and keep it.
- Think before you speak. Once you say something, it is difficult—if not impossible—to take it back. Often you end up regretting words uttered impulsively, no matter how satisfying it was to say them at the time.
- Don't try to "help" someone by softening a hard truth or stretching a welcome truth. You don't really help people by stretching or softening the facts of a situation, and you will probably lose their trust when they eventually learn that the facts you gave them were not accurate.
- Don't tell people what you think they want to hear just to gain favor with them unless what you are telling them is the truth.
- When people ask your opinion, give it tactfully but honestly. Don't wait to see what they think first and then just parrot their opinion. When someone asks your opinion, assume they really want it. Be tactful, honest, and accurate.

Apply these strategies consistently in your conversations with others and you will win the trust of the people you need to communicate with. This does not mean there will not be problems. When you get into the habit of telling the truth, you are going to occasionally rub someone wrong because he really does not want to hear the truth. However, in the long run, the trust you will earn from telling the truth will outweigh any negatives that might come from rubbing an insecure person the wrong way. In fact, even those who don't want to hear unwelcome information or a hard truth will respect you for telling the truth.

I once worked with a technician I will call Hector who was known as a "straight shooter" because he always told the truth when communicating with his teammates. In fact, Hector was so upfront when communicating that he occasionally made supervisors and teammates uncomfortable. Hector simply refused to soften a hard truth or shy away from unwelcome information during team meetings or conversations.

In team meetings, if our supervisor appeared to be withholding information, shading the truth, or stretching the facts, Hector would tactfully and respectfully ask questions for clarification. During one meeting, our supervisor mentioned that our company had won a new contract that had a tight deadline. The supervisor was warning us to expect some overtime during the next 18 months.

While the rest of us shifted uncomfortably in our seats and wondered exactly what the supervisor meant by "some overtime," Hector spoke up and asked a simple question: "Sir, would you tell us exactly what you mean by some overtime?" Clearly, the question made our supervisor uncomfortable because it forced him to reveal some facts he was holding back—facts he knew we did not want to hear.

Before the meeting was over, we knew that our team was facing 18 months of 10-hour days, six days a week. On the one hand, this was unwelcome news—hence our supervisor's reluctance to tell us the whole truth. On the other hand, this was information we needed to hear even if we did not want to hear it. We needed to hear it because working 10-hour days six days a week would affect our lives in significant ways.

Some of us were going to college at night and on weekends. Others had family obligations, hobbies, and even part-time jobs. The new overtime schedule would interfere with all of these after-work obligations. This was unwelcome but essential information for the members of our team. Consequently, we appreciated Hector's efforts to bring out full and accurate information so that we could plan to make the necessary adjustments in our lives. In the long run, Hector's honesty and commitment to accuracy earned him the trust of his teammates and his supervisor. This trust eventually helped Hector secure promotions, and he enjoyed a very successful career.

Listening as a Communication Tool

Hearing is a natural process, but listening is not. A person with highly sensitive hearing abilities can be a poor listener. Conversely, a person with impaired hearing can be an excellent listener. Hearing is the process of physically decoding sound waves, but effective listening requires perception. I define listening as follows:

Listening is receiving a message, correctly decoding it, and accurately perceiving what is meant by it.

Practices That Can Lead to Inaccurate Perceptions

Listening breaks down when the receiver does not accurately perceive the message. In addition to the inhibitors of effective listening explained in the previous section, there are also several practices that can cause you to inaccurately perceive a message or miss it altogether:

- Forgetting to concentrate
- Thinking ahead
- Interrupting the speaker
- Tuning out

To perceive a message accurately, you must concentrate on what is being said as well as how it is being said. An important part of effective listening is properly interpreting nonverbal cues (covered later in this chapter).

Forgetting to concentrate

Concentration requires you to eliminate as many distractions as possible and to focus all of your attention on what is being said—verbally and nonverbally. When you are listening to people speak, look them in the eye, listen carefully, watch for nonverbal cues, and concentrate. Give people a chance to convey their message. Do not get in a hurry or become impatient and start jumping ahead to where you think they are going.

Thinking ahead

Thinking ahead is typically a response to being hurried or just impatient, and this is understandable. After all, people on the job are busy, and there will be times when you really are in a hurry to get a job done on time. When this is the case, remember that it takes less time to hear someone out and accurately perceive their message than it does to start over after jumping ahead to the wrong conclusion and missing the point.

Interrupting the speaker

Interrupting the speaker not only inhibits effective listening, but it can also frustrate the speaker. If clarification is needed during a conversation, make a mental note of it, and wait for the speaker to reach a stopping point. Mental notes are preferable to written notes. The act of writing can distract the speaker or cause you to miss the point. If you find it necessary to make written notes when listening to someone speak, keep them short.

Tuning out

Tuning out also inhibits effective listening. Some people become skilled at using body language that makes it appear they are listening when in fact their mind is focused elsewhere. You should avoid the temptation to engage in such ploys. An astute speaker may ask you to repeat what he or she just said. If your mind is so focused on something else that you cannot pay attention, ask the speaker if you can talk with her at another time.

Practices That Can Improve Your Listening

Most people are not good listeners, but with practice, most can become good listeners. You can improve your listening by practicing the following strategies:

- Remove all distractions.
- Look directly at the speaker.
- Concentrate on what is being said.

- Watch for nonverbal cues.
- Be patient and wait.
- Ask clarifying questions.
- Paraphrase and repeat what the speaker has said.

Remove all distractions

Have you ever tried to talk with someone while he was busy doing something else? If so, you know that you did not have this person's undivided attention. Trying to continue working on a project, complete paperwork, check your e-mail, or do anything else when someone is talking to you is not just distracting to the speaker, it can cause you to misperceive the message or miss it altogether. When someone is talking to you, put aside your work and any other distractions and give them your undivided attention.

Look directly at the speaker

When people speak, they communicate with more than just their words. By looking directly at people who are speaking to you, it is possible to read messages in their eyes and to pick up on other nonverbal cues. Consequently, when listening to someone speak, look directly at her and listen with your ears and your eyes.

Concentrate on what is being said

When listening to someone, concentrating on what is being said can mean the difference between an accurate perception of the message and a misperception. You can waste a lot of time following up on the wrong message if you fail to concentrate during a conversation. For the brief period of time that it takes someone to give you a verbal message, block out your surroundings and other mental concerns and concentrate. In the long run, this practice will actually save you time.

Watch for nonverbal cues

Nonverbal cues include body language, tone of voice, rate of speech, and proximity (how close or how far away someone is when talking to you). Reading these cues is an important part of listening. Nonverbal communication is explained in more detail later in this chapter.

Be patient and wait

Some people have trouble listening when the speaker cannot quite get the message out or is struggling with what to say or how to say it. The tendency is to "help" the speaker by completing the speaker's sentences or putting words in the speaker's mouth. This can be a mistake. Psychologists get some of the best, most useful information from their clients by patiently waiting rather than jumping in to "rescue" them when the words do not come easily. Take a lesson from psychologists. When the person speaking to you is having trouble getting the message out, be patient and wait. If you rush in to help them, you are more likely to cut off the message than receive it.

Ask clarifying questions

When someone is giving you a spoken message, he or she often wants you to do something. When this is the case, it is important to understand fully and accurately what the speaker wants you to do. If you do not fully understand what is being asked of you or are confused about any aspect of the conversation, ask clarifying questions. Never assume anything, and never let the speaker go away without clarifying precisely what he wants of you. This will be especially important for electronics technicians who work directly with customers.

Paraphrase and repeat what the speaker has said

As was explained in the previous paragraph, it is sometimes necessary to ask clarifying questions when listening to someone convey a message to you. Something you should get into the habit of doing all the time, even when clarification is unnecessary, is paraphrasing and repeating back to the speaker to verify and validate your perceptions of what was said to you. This amounts to summarizing what you think you just heard and repeating it back to the speaker. In this way, if your perception is not accurate, the speaker can correct it before you waste time dealing with the wrong problem or doing something different from what the speaker actually requested. If your perception is accurate, it will show the speaker that you listened. This listening strategy can be especially helpful to electronics technicians when receiving instructions concerning some aspect of the job.

A technician I will call Juan learned the value of paraphrasing and repeating back to the speaker the hard way—by not doing it and making a major mistake. Juan worked as an electronics technician for a company that serviced the electronic controls for commercial dry cleaning equipment. One of his company's biggest and best customers placed a service call that was taken by Juan's supervisor. The controls on the customer's biggest dry cleaning machine had malfunctioned and the dry cleaning company had fallen way behind schedule. The company needed a new control box installed right away. Juan's supervisor told him to answer the service call and explained which control box would be needed.

Juan was not sure he heard the inventory number correctly, but he could tell that his supervisor was in a hurry to get the service call completed. Consequently, rather than repeat his instructions back to the supervisor to verify the inventory number, Juan just selected the control box he thought would be the right one. Unfortunately, it wasn't. When Juan showed up at the dry cleaning company with the wrong control box, the customer was understandably upset. Now all Juan could do was drive all the way back to his company, retrieve the correct control box, and once again make the long drive to the dry cleaning company. Because he failed to take a few seconds to paraphrase and repeat his instructions back to his supervisor, Juan turned a two-hour service call into a half-day service call and a frustrated customer into an angry customer.

Listening can be good for your career. It certainly helped a technician I will call Manny build a winning career. After completing his education in electronics technology, Manny went to work for a company that renovated and refurbished military aircraft. This company received contracts from the U.S. Department of Defense to update the electronics systems in military fighters, bombers, and cargo aircraft for the U.S. Navy and the U.S. Air Force. Manny and his team were responsible for installing and testing new electronics systems and for troubleshooting and correcting any problems identified by the tests.

Because its customers were the U.S. Navy and U.S. Air Force, Manny's company had a "zero defects" philosophy. Manny had a supervisor who liked to give instructions only once. Consequently, it was important for members of his team to listen closely. Unfortunately, not everyone did. But Manny did. Whenever his supervisor gave instructions relating to a job, Manny listened carefully. He blocked out everything else and focused on what the supervisor was saying. When the supervisor finished his instructions, Manny always asked clarifying questions if there was anything he did not understand. In addition, before the meeting broke up, Manny would say: "Let me make sure I understand what you just said." Then he would paraphrase the supervisor's instructions.

Occasionally, Manny's teammates would become impatient with him. Several were known to say such things as, "Come on Manny. Stop asking questions so we can get to work." Manny did not like making his teammates angry, but he knew that just one small error on their part could cause a multimillion dollar airplane to malfunction with disastrous results. Consequently, in spite of their prodding to the contrary, he continued to ask clarifying questions and to paraphrase instructions. Finally, Manny's approach became accepted practice in his team—an approach that his teammates begrudgingly accepted.

After Manny had been a member of the installation/troubleshooting team for one year, his team had the lowest rework rate in the company. The company's vice-president for production noticed. As a result, when a department manager position became available, Manny's supervisor—the leader of the installation/troubleshooting team—was promoted. When the vice-president for production asked Manny's supervisor who should replace him as supervisor of the installation/troubleshooting team, he quickly recommended Manny. When asked why he recommended Manny, the supervisor said: "Because Manny is the real reason our team has the lowest rework rate. He's a good technician, but he's an even better listener."

Nonverbal Communication

Nonverbal communication is one of the least understood but most powerful modes of communication. Nonverbal messages often convey more than verbal messages. Consequently, it is important that you learn how to read them. Nonverbal communication is sometimes called "body language," a characterization that is only partially accurate. Nonverbal communication does include body language, but body language is only a part of nonverbal communication. There are actually three components: body language, voice factors, and proximity.

Body Language

A person's posture, hand and arm movements, facial expressions, gestures, and dress—all part of his body language—can convey a variety of messages. Even such things as makeup or the lack of it, well-groomed or unkempt hair, and clean or scruffy shoes can convey a message. You should be attentive to both your body language and that of others. Does your body language say "I am a confident, competent technician who cares about doing a good job" or does it say "I lack confidence and don't care?"

One of the keys to understanding body language lies in the concept of consistency. Is the body language consistent with the message you want to convey? It should be. Is the body language of the person speaking to you consistent with her words? For example, the body language of a technician named Cindy often says: "I am not interested and I don't care." Every time her supervisor calls a team meeting, Cindy slouches in her chair and looks around the room with an expression of boredom. Her work uniform is typically wrinkled and dirty, and she likes to wear her company hat sideways. She is a highly skilled technician, but her nonverbal language is hurting her chances for promotions because it makes her supervisor think she does not care about the job in spite of her verbal claims to the contrary.

Erect posture conveys confidence. Maintaining eye contact and a positive facial expression when talking or listening conveys interest. Proper attention to dress and personal appearance say "I care." No matter what you say verbally, your body language will speak louder. Remember to be attentive to the body language of people who speak to you as well as to your own. They key is consistency between body language and verbal messages. When there is inconsistency, most people will believe the body language.

Voice Factors

Voice factors such as volume, pitch, tone, and rate of speech are also important elements of nonverbal communication. These factors can reveal feelings of anger, fear, impatience, uncertainty, interest, acceptance, confidence, and so on.

As with body language, it is important to look for consistency when making note of voice factors. It is also important to look for groups of nonverbal cues. In other words, do not rely on voice factors alone. Look to see if the speaker's body language matches the tone, pitch, and volume of his voice. Nonverbal messages tend to verify each other if you look for them in groups. For example, if you notice that the speaker's voice is higher-pitched than usual and quivering a little, you might conclude that he is nervous or frightened. If you also notice that he is biting his nails, continually shifting in his seat, and has a pleading sort of look on his face, you are probably right that he is nervous, frightened, or both. The voice factors and the body language are consistent with nervousness and fright.

Use voice factors and body language to validate or invalidate the speaker's words. When the words and the nonverbal cues seem to agree, you are more likely to be receiving a truthful message. When they disagree, you need to ask clarifying questions to get to the bottom of the inconsistency. For example, an electronics technician I will call Alex could sense that something was bothering his teammate, Wen Ho Lee. When he asked Lee if there was a problem, his friend smiled and said: "No, I'm fine." But Alex noticed that Lee's body language and voice factors were inconsistent with his answer. Clearly, Lee was nervous. He could not stand still, did not seem to be able to focus on his work, and was constantly checking his cell phone to see if he had missed a call.

Finally, when Alex saw Lee commit what would have been a serious error if it had gone undetected, he pulled his teammate aside and said: "I know something is bothering you. I've never seen you make a careless mistake like this before. What is the problem?" It turned out that Lee and his wife had needed to rush their baby son to the emergency room early that morning. His wife was at the hospital with their son, and Lee was waiting to hear from her. Because he was a new employee, Lee did not want to take time off from work, but he was understandably concerned about his son. When Alex related the problem to their supervisor, Lee was given the day off and told to get to the hospital as fast as he could. Because Alex paid attention to nonverbal cues, his teammate was able to be at the hospital when his son came out of surgery.

Proximity

Proximity as it relates to nonverbal communication has to do with where you place yourself in relation to those you are communicating with and where they place themselves. When talking with someone about work-related issues, most people are comfortable being about one arm's length away from the other person. Any closer can be interpreted as getting personal and any farther away can be interpreted as being distant. Learn to place yourself according to the message you want to send and to notice the proximity of those who talk to you. For example, to promote communication when you are speaking, stand or sit about an arm's length away from the listener. Also, avoid having obstacles between you and the listener or speaker. For example, rather than sit behind a desk or work table, it is better to come around the desk and sit next to the speaker or listener so there is no obstruction between you.

An electronics technician I'll call Keedra had risen to the level of production manager in her company. As production manager, Keedra was responsible for all the electronics technicians in her department—about 150 people. Whenever technicians would come to her office to talk, Keedra always came around her desk and sat next them. She had two chairs placed in front of her desk facing each other—one for her and one for her visitors. The chairs were arranged about one arm's length apart in distance. In this way, Keedra could listen to the reports, recommendations, and complaints of her personnel with no obstructions between them to inhibit communication.

Keedra learned this strategy when she had been working as an electronics technician for less than a year. She had a supervisor who liked to call technicians into his office to discuss their work. He had a large desk that sat on a platform that was slightly elevated above the level of the floor. Whenever a technician visited his office, this supervisor stayed in his chair behind the desk where he could look down at the person in front of him. Keedra remembered feeling like a child who had been called into the principal's office every time she had to talk with this supervisor. She vowed that if she was ever a supervisor herself, she would use proximity to promote effective communication rather than using it to promote an air of superiority. All these years later, she is making good on that vow, and the technicians who report to her appreciate it.

Verbal Communication

Verbal communication ranks close in importance to listening. Most of the communicating you do on the job will be done verbally. Consequently, it is important for you to be an effective verbal communicator. You can improve your verbal communication skills by applying the following strategies:

- *Show interest.* When speaking with people, show some interest in your topic. If you speak about your topic in a bored, distracted, or nonchalant way, the listener is going to think: "If he is not interested in this topic, why should I be?" Sometimes people will downplay their interest in a suggestion or recommendation they want to make because they are not sure how it will be received. In an attempt to spare themselves the embarrassment of rejection, people sometimes act as if they are not really interested or do not really care about what they are say-ing. This is a mistake. Often the other person's reaction to your sugges-tions and recommendations will be colored by your level of interest and enthusiasm. Remember, if you do not appear to care about your own ideas, why should anyone else?

- *Project a positive attitude.* A positive, friendly, enthusiastic attitude en-hances verbal communication. A caustic, superior, disinterested, or argumentative attitude will shut off communication. Be patient, be friendly, and smile. Remember that there is a natural human tendency for other people to reflect your attitude back to you. If yours is a nega-tive attitude, then a negative attitude is what will be reflected back to you. A positive attitude when conveying a verbal message is more likely to be received positively.

- *Be flexible.* There is usually more than one way to solve a problem or im-prove performance on the job. When you are verbally presenting your ideas, suggestions, and recommendations to others, be flexible. Your way is seldom going to be the only way. Be open to considering the opinions and thoughts of other people. Ideas that are presented dogmatically are not likely to be well received by others. I know of an electronics technician who has excellent technical skills but never seems to be able to get a promotion. One of the reasons for his failure to climb the career ladder is his inflexibility. He has a habit of present-ing an idea and then attacking anyone who offers a counter proposal or suggests even a minor revision to his proposal. Consequently, his ideas are rarely accepted, and his opinion is rarely solicited.

- *Be tactful.* Tact is an important ingredient in verbal communication, particularly when delivering a sensitive or potentially controversial message. Tact has been referred to as the ability to make your point without making an enemy. The key to tact in verbal communication is to think before you talk. Consider not just what you want to say, but how you want to say it. Consider how what you say and how you say it will be received. The same verbal message said one way will be re-ceived differently than if said another way. It is especially important to be tactful when disagreeing with someone or correcting someone. As-sume you are the team leader in a team of electronics technicians. You need to give one of your team members—Nhieu Vo Minh—corrective feedback. One way to give the feedback would be to approach Vo Minh and say: "You made a mess of that job I asked you to do. Do it over and see if you can get it right this time." Another way to give the same feedback would be to say: "That last job I asked you to do did not come up to your usual high standards of performance. Let's take a look at it and see if I can help you correct it." The first approach makes the point, but it does so in a way that would be either hurtful, insulting, or both. The second approach makes the same point but without being hurtful or insulting. Which do you think would be received better by Nhieu Vo Minh? Which approach would you prefer if you were in Vo Minh's place?

- *Be courteous.* Being courteous means showing appropriate respect for the receiver of your verbal messages. When you want to deliver a verbal message, it is courteous to make sure you are not interrupting the other person or that you are approaching her at a good time. Courtesy is the act of taking into consideration the other person's needs, and it can have a lot to do with how your verbal message will be received. For example, stopping a teammate just as she is rushing out the door at the end of a long workday and trying to give her a verbal message is not courteous—nor is it smart. If she is in a hurry to get home or anywhere else, she is not going to be open to receiving your message. Ignoring her needs in this situation shows a lack of courtesy, which is likely to result in a lack of interest on her part. To make sure that your verbal messages are received in a positive way, remember to be courteous to the receiver.

Review Questions

1. Are people born good communicators, or can they learn to be good communicators? Explain.
2. Distinguish between communication and effective communication.
3. List and explain the various components of communication.
4. List and explain common inhibitors of communication.
5. Define in your own words the term *listening*.
6. List and explain common inhibitors of effective listening.
7. Explain several strategies you can use for improving your listening.
8. List and explain the three components of nonverbal communication.
9. Explain how you can improve your own verbal communication.

Discussion Questions

1. Discuss with other members of your class the following statement: "As long as I have good technical skills, I don't need to be a good communicator." Do you agree or disagree? Why?
2. Discuss with other members of your class the following statement: "One person communicates in the same way as another. I don't worry about inhibitors. I tell people what I expect. It's their problem to figure out what I mean." Do you agree or disagree with this individual? Explain your reasoning.

3. Discuss with other members of your class the following statement: "I don't worry about listening. I have two ears, and I can hear well. I'm sure I'm a good listener." Do you think this individual is a good listener? Why or why not?

4. Discuss among your classmates the issue of nonverbal communication. Explain how you can tell if a person is angry, sad, happy, or bored by reading nonverbal cues.

5. Discuss your response to the following question with your class: I don't seem to be able to get people to understand what I am saying—how can I improve my verbal communication?

CHAPTER 8

THINK CRITICALLY TO CONTINUALLY IMPROVE PERFORMANCE

LEARNING OBJECTIVES

Upon completion of this chapter, you should be able to do the following:

- *Define the term* critical thinking *as it relates to electronics technicians.*
- *Explain why it is so important for organizations and their personnel to continually improve performance.*
- *Explain the role of critical thinking in the continual improvement process.*
- *Summarize the benefits of critical thinking.*
- *Describe how you can become a critical thinker.*

The most successful electronics technicians are always looking for ways to do their jobs better. By better, I mean more efficiently and more effectively. This is how they become productive, and employers need productive electronics technicians. The need to get better and better all the time is the concept of *continual improvement*. Critical thinking is fundamental to continual improvement, and continual improvement is fundamental to success in a competitive environment. This chapter shows you how to be a critical thinker who helps his or her organization continually improve.

What Is Critical Thinking?

Critical thinking as it relates to electronics technicians is an approach to doing your job that involves objectively evaluating situations, directions, problems, processes, opinions, arguments, and recommendations rather than just taking them at face value or making assumptions about them. A critical thinker never accepts that a procedure is right or a process is appropriate just because that's the way it's always been done. Rather, critical thinkers are objective evaluators of the world around them who make decisions and form opinions based on facts and the best analysis of those facts they can conduct in the time available to them.

Why Continual Improvement Is So Important

Employers of electronics technicians operate in a competitive environment. Competing in business is like competing in the Olympics in that performance that was good enough to win in the last games will not be in the next games. Olympic athletes continually improve their performance and so do business competitors. This is why continual improvement is so important. Correspondingly, it is why critical thinking is so important.

Critical thinkers never perform their jobs the way they do simply because that is way things have always been done. Rather, critical thinkers understand that the only acceptable way to perform their jobs is the best way; meaning the most efficient and effective way that will result in the highest quality. Consequently, critical thinkers are always looking for better ways to do their jobs. When personnel in organizations think and work this way, performance improves continually, and the organization becomes more and more competitive.

Role of Critical Thinking in Continual Improvement

The organization where you work after graduation—regardless of whether it is large, small, public, or private—will operate in a competitive environment that undergoes constant change. In fact, the electronics industry changes faster than almost any other. To compete, your employer must be able to keep up. For example, just think about cell phones.

The original market leader in the manufacture of cell phones was Motorola. Its cell phones were based on analog electronics technology, and they were so good that Motorola was the leading manufacturer in the world of this high-demand product. That is, they were the world leader until they became complacent and stopped improving their technology.

Decision makers at Motorola were slow to realize that the world of electronics was transitioning from analog to digital technology. This oversight allowed a small company few people had ever heard of at the time—Nokia—to introduce digital cell phones and quickly become a big company everybody has heard of. Before long, other companies began to follow Nokia's lead and cell phones based on digital technology became common. Motorola was forced to watch from the sidelines as its market share rapidly dwindled.

Motorola eventually decided to "climb on board the digital train," but by that time, there were numerous other companies already on it and with better seats closer to the front than Motorola. Motorola's line of digital cell phones eventually allowed the company to regain some of its lost market share, but the company has not and probably will not ever gain the market dominance it enjoyed during the old days of analog technology.

What does the concept of market share mean to you as an electronics technician? The short answer is that it means jobs and wages. The more market share an organization has, the more employees it needs and the more it can pay those employees. For example, you would not want to be an electronics technician at a company that lost a major share of its market like Motorola did in the earlier example because lost market share can mean the company is forced to lay off, terminate, and retire employees while freezing or even lowering the salaries and benefits of those they keep.

Gaining and keeping market share in a competitive environment is about staying competitive, and staying competitive is about continually improving performance. Critical thinking is essential to continual improvement because continually improving the performance of people, processes, and products requires critical thinking.

An organization's overall performance is largely the sum total of the performance of executives, managers, supervisors, and employees. When Motorola was slow to see the trend away from analog electronics to digital in the cell phone industry, its key decision makers were not thinking critically. They appear to have thought that because they were the market leader at the time, they would always be the market leader. This is the opposite of critical thinking.

An organization competing in the global marketplace is like a football team competing in the Super Bowl. To win the game, it needs all of its personnel—coaches and players—giving their best and thinking critically about how to do better on every play. Players on a Super Bowl team succeed by playing hard and playing smart. Playing smart requires critical thinking. This is also true of electronics technicians. To succeed individually and help their employer succeed too, electronic technicians need to work hard and work smart. Working smart means being a critical thinker.

Potential Benefits of Critical Thinking

Critical thinkers approach their work in much the same way as a detective. Whereas a detective looks and listens objectively with a mind to finding the truth, a critical thinking electronics technician looks and listens with a mind to finding the best work practices. What follows are some specific benefits you can gain from being a critical thinker in your job as an electronics technician:

Ability to Identify Best Practices

To achieve peak performance, you have to identify and apply what is known in the workplace as *best practices*. With any work-related task, there are best practices that, collectively, represent the best way to perform the task in terms of efficiency, effectiveness, and quality. This is the "working smart" component of the success strategy that says "work hard and work smart."

Electronics technicians who are *not* critical thinkers tend to perform tasks the way they were taught and in the way they have always been done. Technicians who *are* critical thinkers mentally challenge the way things have always been done by regularly asking themselves if there is a better way to do this.

More Value to Team Members

Critical thinkers do more than just make themselves more productive, they also make their teams more productive. This, in turn, makes critical thinkers more valuable to the team. Team performance is evaluated against various standards. Typically, one of these standards is relative performance when compared to other teams in the organization that have the same or similar duties. Any team member who, by thinking critically, helps his or her team outperform other teams will be more valuable to the team and the overall organization. This is why sports teams give "Most Valuable Player" or MVP awards to their best performers each year. When working as an electronics technician, critical thinking will help you be your team's MVP.

Credibility and Prestige

Critical thinkers who are able to find ways to make the team perform better quickly earn credibility with their team members. Over time, this credibility translates into prestige as critical thinkers who help the team perform better gain status in the eyes of their teammates and others in the organization. This is like being the best hitter on a baseball team. Your hitting helps the team win more games, which, in turn, gives you credibility and prestige in the eyes of your teammates, coaches, and managers.

Better Problem Solving

Critical thinking will make you a better problem solver, which is important because problem solving is essential to your success. In fact, it is when attempting to solve problems that critical thinking is most important. Many of the jobs performed by electronics technicians involve troubleshooting, which is just another name for problem solving. Whether the problem in question is a malfunctioning piece of electronics equipment or an ineffective work process, troubleshooting to find the cause of the problem and recommending a solution requires critical thinking. Remember, critical thinking involves maintaining an objective perspective that bases decisions on facts.

More Support for Ideas

When you have an idea for improving a work process or procedure, you cannot expect your supervisor or teammates to automatically accept it just because the idea is yours. You must be prepared to win support for your ideas by logically and factually showing how what you propose is better than what exists. Preparing logical, factual arguments is a function of critical thinking. Critical thinking should be applied not just to the arguments of others but to your own as well. When you critically evaluate your own ideas before presenting them for the consideration of others, you are more likely to win support for them because you will have already thought of the various concerns they might raise and answered them for yourself.

Better Focus During Discussions

Noncritical thinkers tend to lose their focus during discussions and get sidetracked by irrelevant issues that do not relate to the topic at hand.

This is not the case with critical thinkers. Because they concentrate on the heart of the issue, critical thinkers are able to sift through unrelated points that are raised and the mental "fog" they can create to stay focused on what really matters. This ability to stay focused during discussions makes critical thinkers more valuable participants in discussions about how to solve problems and make improvements.

Becoming a Critical Thinker

Consider the following example of two electronics technicians—Debby and Marie—who work for XYZ Electronics, Inc. Debbie is a critical thinker, but Marie is not. As a result, Debbie is enjoying a more successful career than Marie. While doing her job, Debbie is always trying to find ways to improve her performance as well as that of her team. Marie does a good job, gives her best effort every day, and is always on time, but she is uncomfortable with change. If Marie had her way, she would learn how to perform a task and then do it that way forever. She likes doing things the way they have always been done. However, once she has learned how to perform a task, no one does it better.

XYZ Electronics, Inc. has just won a new contract that is the largest in the company's history. That is the good news. The bad news is that to complete the contract on time and under budget, XYZ is going to have to re-engineer its manufacturing processes to achieve greater productivity. Every manufacturing process will have to achieve maximum efficiency and effectiveness. Consequently, XYZ's management personnel are looking for technicians in the company's manufacturing division who are known to be critical thinkers to serve as team leaders.

These new team leaders must be 1) good at their current jobs (dependable, good team players, and talented technicians); and 2) able to find ways to continually improve the performance of people, processes, and procedures. Since being named one of these new team leaders would mean both a promotion and a raise, Marie is hoping to be recommended by her supervisor. However, when the recommendations are made, it is Debbie, not Marie, who becomes the team leader.

Their supervisor considered both Marie and Debbie carefully. At first, he had trouble choosing between the two technicians. Both Debbie and Marie are good team players, dependable, and talented technicians. But when the supervisor thought about the issue of critical thinking, his choice became obvious. Marie fell short on this criterion, whereas Debby excelled. Clearly, Debby was the type of technician XYZ's managers were looking for to lead a manufacturing team in carrying out the company's new contract. Only one characteristic separated Debby and Marie—that characteristic is critical thinking.

Steps to Becoming a Critical Thinker

Critical thinking is a skill you develop with practice and patience, not a talent you are born with. Because critical thinking is a skill, most people can become critical thinkers provided they are willing to put forth the effort. What follows are several steps for becoming a critical thinker.

Practice viewing your work objectively

People tend to see things as they want to see them rather than how they actually are. Critical thinkers, on the other hand, see things as they are, not as they would like them to be. The difference in these two perspectives is objectivity. Viewing your work objectively means looking at it with your head rather than your heart. Critical thinkers view their work objectively, look for facts, and go where the facts lead them. This is exactly what electronics technicians do when they troubleshoot a piece of malfunctioning equipment. Noncritical thinkers, on the other hand, view their work emotionally and go where they want to go in spite of the facts.

For example, I once knew a technician I will call Mike who worked as a printed circuit board designer back in the old days before CAD systems had been invented. In those days, printed circuit boards were laid out using special tape and other adhesive materials. Mike was the best in the company at laying out printed circuit boards in this way. He was fast, accurate, and efficient.

When CAD systems were developed, and design software for electronics applications began to show up at trade shows, engineering managers in Mike's company started to advocate for a transition from the hand-taping method to CAD. Mike tried out a couple of printed circuit board packages but persistently discouraged their adoption. He claimed that "...no computer will ever outperform me in laying out printed circuit boards."

Because Mike's expertise was in manually laying out printed circuit boards and because his status in the company was tied to how well he did this, Mike did not want the design process to change. He had an emotional attachment to it. Consequently, rather than look at the situation objectively as a critical thinker would, he looked at it emotionally grasping at any shred of evidence that supported how he wanted things to be as opposed to how any objective person could see they were going to be.

Other electronics technicians less attached to the status quo than Mike begin to experiment with laying out printed circuit boards on CAD systems. Admittedly, the first systems that came out were not very good, but as time went by, they got better and better, and Mike's more objective teammates got better and better at using them. Within a couple of years, Mike had been bypassed by these critical thinkers. Eventually, even Mike could see the handwriting on the wall and realized he would have to change. He did, but he had fallen way behind the electronics technicians in his company who had objectively moved ahead with CAD.

Learn to mentally ask questions

Remember that nothing improves without changing. You cannot keep doing things the way they have always been done and expect performance to improve. Consequently, as you go about the business of doing your job, learn to mentally ask questions such as these:

- Why do we do this in this way?
- Is there a better way to perform this task?
- Will what is being proposed really make things better?

Asking questions such as these does not mean that you stop doing your work or that you become a pest to your supervisor by constantly asking these questions out loud when you are supposed to be working. Rather, it means that while you are working, you mentally ask yourself these questions. You can also ask yourself these questions outside of work. I keep a notepad on the dashboard of my car, next to my favorite chair in the den, and next to my bed so that when an idea pops into my mind about a better way to do something, I can write it down. Once you begin to mentally challenge the status quo, doing so will become a natural process—you will find yourself doing it without even thinking. It just becomes second nature.

Practice objectivity

One of the most difficult things for a human being to do is look at the world objectively. We all have our own individual perspectives, points of view, and biases. We all have our own views of how we would like things to be. Consequently, it can be difficult to be objective—especially when the issue in question is personal (for example, my work, my idea, my opinion, and so on).

For example, consider the fact that many of the deaths attributed to heart attacks could actually have been prevented if the victims had chosen to view their symptoms objectively and sought early medical intervention. Instead, many people who suffer debilitating heart attacks ignore such symptoms as chest pain, shortness of breath, and numbness in one or more limbs because they do not want to admit that this could be happening to them. They engage in misguided wishful thinking instead of objective critical thinking and, as a result, suffer permanent heart damage or, worse, they die.

Being objective means putting aside personal agendas, feelings, biases, and interests; looking at the facts concerning the situation; and making the decision or taking the action that is indicted by the facts. With practice, objectivity will become easier, but it will never be easy. You will probably never reach a point where you can effortlessly accomplish objectivity about anything that is personal to you, such as your work. Being objective will always require a concerted effort on your part, but you can do it.

Make performance rather than popularity your goal

Objective people never just go along with the crowd unless the facts of the situation indicate that the crowd is headed in the right direction. Some people are reluctant to disagree with others, and so they go along, even when they disagree or see a better way. On the job, it is nice to be popular, but it is essential to be right. When critical thinkers are forced to choose between being right and being popular, they choose to be right.

Critical thinkers learn to view their work from the perspective of improving performance rather than being popular. This is important because the popularity gained by people who make the wrong decision just to be popular is destined to be short-lived anyway. In the long run, the most popular people will be those who help their team perform better.

Learn to analyze arguments carefully

As an electronics technician, you will sometimes be called upon to participate in brainstorming sessions or open discussions about ways to improve performance or solve problems. In such situations, critical thinkers try to be open to other points of view and opinions, but they also learn to mentally analyze them carefully. Figure 8-1 contains several criteria you can use to analyze the arguments made by teammates and other people.

Learn to carefully analyze explanations

When people on the job want you to do something or go along with something they want to do, they will typically explain things in terms that

---

Criteria for
ANALYZING ARGUMENTS

---

1. What are the *issues* being raised?
2. What *opinions* have been stated?
3. Are the statements being made *credible*?
4. Are the statements being made *consistent*?
5. Are the statements being made *relevant*?

FIGURE 8-1 Arguments should be analyzed critically.

Criteria for
ANALYZING EXPLANATIONS

---

1. Does the explanation make sense based on what you know about the situation?
2. What are the alternatives available in this situation?
3. Is the basic premise of the explanation valid?
4. Does the explanation contain any false assumptions?

FIGURE 8-2 Explanations should be analyzed critically for validity.

---

favor their point of view. For this reason, it is important to develop the ability to carefully analyze explanations. Figure 8-2 contains several criteria you can use to analyze the explanations people give you at work.

To become a critical thinker, focus your mental energy on improving performance. Never agree with a teammate's opinion just to win favor or be liked. Be tactful when you disagree, and be open to other points of view. After all, your teammate's idea might be a good one. However, never just go along to get along. If you see a better way, speak up. Be tactful, but have your say. That is what critical thinkers do.

Review Questions

1. Define the concept of *critical thinking.*
2. Why is continual improvement so important in the workplace?
3. Explain the role of critical thinking in the continual improvement of performance.
4. List and explain the potential benefits of critical thinking.
5. How can you become a critical thinker?

Discussion Questions

1. Have you ever been in a situation in which someone became emotionally attached to an argument or opinion when even a small dose of objectivity might have changed his or her mind? Discuss this situation.
2. Why do you think people can get so attached to and comfortable with a certain way of doing things? Are you this way? Discuss.
3. Defend or refute the following statement: Just because electronics technicians learn to be objective troubleshooters does not mean they will automatically be good critical thinkers.

4. Think of a time when someone was trying to convince you to do something or to go along with what they wanted to do. Discuss how they went about presenting their arguments and any flaws in the logic or shortcomings in their reasoning you can remember.

Application Assignments

1. Turn on a television talk show that discusses items in the news (politics, news, world events, and so on). As you listen to the discussion and debate, apply the criteria for analyzing arguments set forth in Figure 8-1.
2. Contact a local elected official in your community. Ask why he or she voted for or against a given issue. As you listen to the explanation, apply the criteria set forth in Figure 8-2.
3. Identify someone who works in electronics and ask this person how critical thinking or the lack of critical thinking comes into play in the workplace (good and bad). Ask for specific examples you can share with your classmates.

CHAPTER 9

LEARN TO WORK EFFECTIVELY IN A DIVERSE ENVIRONMENT

LEARNING OBJECTIVES

Upon completion of this chapter, you should be able to do the following:

- *Define the term* diversity *as it applies to the workplace.*

- *Explain how resistance to diversity is a learned concept.*

- *Explain key diversity-related terms and concepts.*

- *Explain how being prejudiced toward others can harm your career.*

- *Explain several strategies that can help people overcome the prejudices they have learned.*

How would you answer the following question: What do Americans look like? When you think about this question, you soon realize that there is no quick and easy answer to it. Some Americans have dark skin, and others have light skin. Some have straight hair, and others have curly hair. Some Americans have black hair, and others have blonde, brown, red, or a hundred different other shades. Some have round eyes, and others have slanted eyes. Some Americans have blue eyes, and others have green, brown, gray, hazel, or a hundred different other shades. Some Americans are tall, and others are short. Some are thin, and others are thick.

When you think about all of the different races, cultures, backgrounds, and national origins that have produced the American population, the only realistic answer to the question posed earlier is that Americans look like the world. American citizens are descended from virtually every country, race, and culture in the world, making the United States one of the most diverse countries on the planet. It is important to understand this fact because the workplace you will enter after completing your electronics studies will mirror the diversity of the country. Consequently, it is going to be important for you to be able to work effectively in a diverse environment.

This chapter explains why the ability to work in a diverse environment is so important for electronics technicians who want to have a successful career. It also provides *success strategies* that will help you work effectively with people who are different from you in terms of race, culture, national origin, gender, religion, politics, physical characteristics, background, personality, and all of the other ways in which people can be different.

Diversity Defined

Try this experiment. Ask several fellow students, friends, and family members what comes to mind when they hear the term *diversity*. I tried this experiment while writing this chapter and found that most people I asked thought the term had to do with racial differences. In a way it does, but race is just one aspect of the concept of diversity. There are many others. Diversity means the quality of being different. As it relates to your career, diversity is about how people in the workplace are different. Of course, race is one way in which people in the workplace can be different, but there are also many others.

In fact, people of the same race can be quite different from each other. Even people from the same family—brothers and sisters—can be vastly different from each other in terms of such attributes as age, gender, physical ability, intelligence, religious beliefs, education level, personality, skills, experiences, personal preferences, appearance, attitude, political orientation, and so on. If there can be this many differences between family members, imagine how many differences there might be between you and the people you will work with over the course of a career.

One certainty about your career as an electronics technician is that you are going to work with people who do not look, talk, dress, eat, interact, socialize, believe, or react like you. In fact, with some of your fellow electronics technicians, the only things you will appear to have in common, at least on the surface, will be work related. Fortunately, work-related goals, aspirations, ambitions, and practices are all you really need to have in common to work effectively in a diverse setting. In addition, once you establish common ground from the perspective of work, you will typically find that you have more in common with your teammates than you might have thought at first. In fact, people who find themselves in a setting in which the people are different from what they are accustomed to are usually surprised to learn just how much they have in common with them.

Your desire to succeed as an electronics technician makes natural allies of you and others in your organization who also want to succeed, regardless of the many other ways in which you might be different. No one in an organization succeeds alone. Electronics technicians work in teams, and people in teams depend on each other to perform at peak levels—a prerequisite to success. This mutual dependence can help build bridges between people in the workplace who come from different backgrounds, races, points of view, and so on.

The most successful electronics technicians are good team players who have learned to focus not on how they and their teammates are different but on what they have in common (for example, ambition to succeed, need to be a peak performer, desire to continually improve, and so on). Think of a professional baseball team. Few teams are more diverse than modern professional baseball teams. Because baseball has become an international sport and because playing in the American major leagues represents the pinnacle of success to baseball players, professional baseball teams in this country attract people from all over the world. Watch almost any professional baseball game on television and you will see players from the United States, Japan, and a variety of Latin American countries—all on the same team. You will see African-American, Caucasian, and Latino players, all playing together to achieve a common goal—winning the league championship and going to the World Series.

When an African-American player steps up to the plate, all other members of his team—those of Caucasian, Asian, and Latino backgrounds—are pulling for him to get a hit. The hitter's race or country of origin do not matter to his teammates. All that matters is that he succeed as a hitter because his success at the plate contributes to the team's success, which, in turn, contributes to the success of the other individual players. Hence, the desire to succeed in your work team can be all the common ground that is needed to begin building bridges between people of different backgrounds. When this happens, you will begin to find that people who appear to be different on the outside can be very similar on the inside.

Diversity-Resistance Is a Learned Behavior

If the most successful electronics technicians are good team players, and if teams in today's workplaces are typically diverse in their composition, it stands to reason that technicians would want to learn how to work well in a diverse environment. Many do. However, in spite of the benefits of working well in a diverse setting, some people still resist learning how. People who are resistant to working in a diverse environment are not necessarily bad people. They are just people who have learned some bad behaviors. The term *learned* is important here because most of the negative behaviors displayed by people who resist diversity are learned behaviors.

To understand the concept of learning to be uncomfortable around or resistant to people who appear to be different, consider the example of young children. Left to themselves, little children will happily play with each other without a thought to race, gender, or other differences. They are just happy to have playmates. It is only as they grow older that they *learn* to adopt negative attitudes and behaviors toward people who are different from them.

These learned negative attitudes and behaviors include prejudice, labeling, stereotyping, and inflexibility in opinions, perspectives, and points of view. It is important to understand that these attitudes and behaviors are learned because what can be learned can be unlearned. Said another way, people who can learn negative behaviors can also learn positive behaviors. I saw a powerful example of this while serving in the U.S. Marine Corps.

I knew two Marines who, on the surface, appeared to be as different as two people could be. One was an African-American Marine who grew up in a large city in the north. The other was a Caucasian Marine who grew up on a farm in the rural south. When they were thrown together in boot camp, they were deeply suspicious and distrustful of each other. Each had learned to be prejudiced toward the other's group. The African-American Marine had learned, erroneously, that all people from the rural south hated blacks and were members of the Ku Klux Klan. The Caucasian Marine had learned, erroneously, that all inner-city blacks hated whites and were members of drug gangs. These erroneous prejudices got these two individuals off to a bad start that began with insults, escalated into arguments, and culminated in a fight.

Although to someone unfamiliar with the Marine Corps' methods, this situation might appear to have been a major problem, in fact it was an excellent opportunity for both of these individuals to unlearn the inaccurate assumptions that led to their illogical prejudices. Because the Corps knows that people from different backgrounds may have to depend on each other in a life-or-death combat setting, the Marine Corps has developed a unique and effective way of handling these types of situations. Rather than separate these two recruits to keep them apart, the Marine Corps applied a time-tested strategy. It forced them to work together. It created a situation in which they could not succeed without helping and depending on each other. Using the Marine Corps' own unique version of the "buddy system," the drill instructors in our recruit platoon made these two recruits eat together, bunk together, march together, exercise together, run the obstacle course together, and solve field problems together. If one made a mistake, both were punished. If one had to rerun the obstacle course, both had to rerun it.

Before long, these two former enemies learned that they had at least one thing in common—the need to work together to survive the rigors of boot camp on Parris Island. Once this realization dawned on them, they began to learn that they also had other things in common. Both were excellent shots on the rifle range, and both excelled at the physical aspects of boot camp. In addition, they both liked to march and were good at close order drill. By the time our platoon graduated from boot camp, these two former enemies—one a street-smart, inner-city African-American from the north and the other a high-school dropout and farmer from the rural south—were inseparable friends. They had learned a valuable lesson that I am sure served them well throughout their lives: that people who appear on the surface to be vastly different can actually have more similarities than differences.

Diversity-Related Terms

The concept of diversity in the workplace has a language all its own. This language consists of several terms representing concepts that all electronics technicians who will work in a diverse setting should be familiar with. These terms and concepts are prejudice, stereotyping, discrimination, and labeling, (see Figure 9-1).

Prejudice, stereotyping, discrimination, and labeling are inherently illogical practices that have no place on the job or in any other aspect of our lives. The overriding goal of any organization that employs electronic technicians is success in the marketplace. Success in the marketplace is achieved in large measure by consistent peak performance and continual improvement. Prejudice, stereotyping, discrimination, and labeling can

DIVERSITY-RELATED CONCEPTS

- Prejudice
- Stereotyping
- Discrimination
- Labeling

FIGURE 9-1 Electronics technicians should understand these concepts.

rob an organization of the energy, vitality, and teamwork needed to be globally competitive. Organizations that allow prejudice, stereotyping, discrimination, and labeling to persist are not likely to rise above the level of mediocrity in the long run, and in a competitive environment, mediocre organizations eventually fail.

Prejudice

Prejudice is an internal predisposition to adopt generalized negative perceptions about other groups of people. People who are prejudiced toward other groups of people observe what they believe to be a negative attribute in one member of a group and then attribute that negative attribute to everyone in the group. This is like observing that a certain Asian-American student has difficulty with mathematics and then generalizing this observation to mean that all Asian-American students have difficulty with mathematics.

Prejudice reveals itself in the workplace in a number of negative ways, including discrimination, stereotyping, and labeling to name just a few. People who are prejudiced against another group of people feel that they are superior in some way to those people. A male supervisor who is prejudiced against women might discriminate against female electronics technicians when making recommendations for promotions, bonuses, or raises. A female supervisor who is prejudiced against men might stereotype them by saying all men are too clumsy to be printed circuit board assemblers. A supervisor who is prejudiced against people of another race might label them as all being lazy or incompetent.

People who are prejudiced divide all of the people in the world into "us" and "them." Those who are like these prejudiced people—according to whatever characteristics they choose to apply—fall into the "us" category. Everyone else falls into the "them" category. Those who fall into the "us" category are considered to have positive characteristics based on nothing more than their status of being the *same*. Those who fall into the "them" category are considered to have negative characteristics based on nothing more than their status of being *different.*

Prejudice-related experiment

Prejudice can be incredibly arbitrary in its applications. People can be different in a lot of ways. For example, take a moment and make a list of all the ways you can think of that people might be different. Here is just a brief list that I developed in less than five minutes:

- Mental ability
- Physical ability
- Physical appearance
- Age
- Marital status
- Geographical location
- Religion
- Ethnic background
- Nationality
- Education level
- Opinions
- Values
- Political beliefs
- Interests
- Personality
- Cultural background
- Taste in music
- Taste in food
- Height (tall or short)
- Career (blue collar or white collar)

Now that you have your list, try this. For each item on your list, identify a prejudiced attitude you are familiar with or aware of that is associated with that item. To get you started, several examples follow:

- *Mental ability.* Prejudiced myth: All athletes are dumb. Since it takes a certain amount of intelligence to play any sport, this prejudiced belief is obviously illogical. But it is more than illogical, it is also false. Some of the most intelligent people in the world are, or have been, athletes.
- *Geographical location.* Prejudiced myth: All people from San Francisco are liberal democrats. Many are. San Francisco enjoys its reputation for being a liberal city that leans strongly toward the Democratic party. However, even in San Francisco there are moderates and conservatives.
- *Religion.* Prejudiced myth: All Christians hate Moslems and vice versa. In reality, Christians and Moslems in the United States have an admirable record of peaceful accord in which members of both religions coexist in harmony.

- *Political beliefs*. Prejudiced myth: All conservatives are rich business-people who do not care about the poor. In fact, only a minority of conservatives are wealthy. More are actually middle-income people who have an admirable record of charitable work on behalf of the poor.
- *Cultural background*. Prejudiced myth: All African-Americans like rap music. Some do, and some do not. Some like jazz, some like classical, some like contemporary, and some like country music. In fact, more rap CDs are purchased by Caucasians each year than by African-Americans.
- *Taste in food*. Prejudiced myth: All Latinos like hot, spicy food. Some do, and some do not. Taste in food is a personal, individual concept—not one that can be attributed to a group.

You can see from just these few examples how illogical prejudiced thinking can be. There is really is no characteristic that can be accurately attributed to any group of people. The minute you begin to think that all people in a given group share a certain characteristic, you need to step back and consider how illogical such a thought is. Although it may be true that *some* members of a group share a certain characteristic, it will be equally true that *some* do not. Further, it will also be true that *some* members of other groups also share the characteristic in question.

Prejudice and the self-fulfilling prophecy·

One of the most damaging and disturbing aspects of prejudiced thinking is what is known as the *self-fulfilling prophecy*. This is a psychological concept that can work for either good or bad depending on its application. Unfortunately, when coupled with prejudiced thinking, its application can be decidedly negative. A supervisor who thinks, for example, that women cannot perform mechanical jobs that require the use of hand tools might turn this prejudiced myth into an actually reality by never giving them the chance.

People in positions of authority or influence (parents, teachers, professors, supervisors, and so on) who expound often enough on their prejudices might actually convince those they are prejudiced against that they are right. For example, an influential teacher—particularly one who teaches young people—might actually convince the boys in his class that they should not become chefs because cooking is "woman's work." Another influential teacher might convince the girls in her class that they should not pursue a military career because military service is for men. Prejudiced myths such as these can become self-fulfilling prophecies when espoused by influential people.

Stereotyping

Stereotyping is one of the negative by-products of prejudice. It is the act of generalizing the characteristics of individuals to whole groups. For example, have you ever heard a "dumb blonde" joke. Such jokes are a form of stereotyping because they falsely imply that all blonde people are dumb. In truth, some blonde people are intelligent, and some are not—a fact that also applies to people with black, brown, or red hair, as well as people who have no hair.

Over the years, I have seen many different stereotypes fall by the wayside as the people they are directed at prove them wrong. When I was young and still playing football, there was a stereotype that African-American athletes were incapable of playing quarterback. As a consequence of this stereotype, many outstanding high school and college quarterbacks who were African-Americans were forced to switch to running back, receiver, or defensive back in order to continue their football careers at the next level.

I find that this stereotype is difficult for the young people of today to believe since some of the best quarterbacks who ever played the game have been African-Americans. Just watch college or professional football on any given weekend, and you will find that gifted black athletes have proved how illogical and wrong this stereotype was, but the stereotype persisted for many years before finally being eliminated by the performance of black athletes on the playing field.

Discrimination

Discrimination is prejudice and stereotyping put into action. It involves treating an individual or a group unequally on the basis of some diversity-related characteristic (race, gender, physical status, culture, or any other way that people can be different). Think back to the issue of African-American quarterbacks in college and professional football mentioned in the previous paragraphs. In the days when African-American athletes were denied opportunities to play quarterback for no reason but race, coaches were going beyond just stereotyping to actually practicing discrimination.

To *think* that a black athlete cannot play quarterback is to engage in stereotyping, but to actually deny such a person the opportunity to prove you wrong is to practice discrimination. As a country, the United States has evolved from a time when some of its citizens openly and unapologetically practiced discrimination on the basis of race, gender, age, and other characteristics to a time when discrimination on the basis of such characteristics is actually against the law.

Equal opportunity is now the law in the workplace, and it is a law that is vigorously enforced by government agencies such as the Equal Employment Opportunity Commission (EEOC). However, discrimination can be a pernicious and persistent practice that surfaces in subtle ways in the workplace. For example, say the members of a team of electronics technicians are all more than 40 years old except one—a 21-year-old technician who just graduated.

The older members of this team take their lunch together every day. During lunch they often discuss work-related issues, problems, and opportunities. They also get together socially once a week. Because they think their younger team member would not fit in (stereotyping), they do not invite him to lunch or their social gatherings (discrimination). This form of discrimination is probably not illegal—although one should not automatically make that assumption—but it is wrong because one team member is being denied potentially advantageous associations with his teammates because of his age. As a result, he is not being accepted as a member of the team and is missing out on important discussions about the team's work.

Labeling

Labeling is an extension of stereotyping that involves attributing a certain characteristic to a group and then using that characteristic to label people in that group. For example, people who are talented in the areas of math, science, or computers are often labeled as "geeks." When this label is used, it is not intended as a compliment.

People with southern accents are sometimes labeled as "rednecks" and thought to be both illiterate and backward. Yet the fallacy of such labels is self evident. Some of this country's most brilliant scholars, poets, musicians, artists, scientists, and political leaders have been men and women who spoke with a southern accent. Several such people have been presidents of the United States.

Why Prejudice Will Harm Your Career

There is an important rule of thumb relating to success in the workplace that all aspiring electronic technicians should understand:

In the long run, the most successful employees are those that contribute the most to the success of their organization.

The organization in question might be a team, unit, department, division, or company. Regardless of the size or composition of the organization in question, if you want to succeed, you must help your organization succeed. This is an unalterable truth in today's globally competitive workplace. Another unalterable truth is that prejudice and its related practices—stereotyping, discrimination, and labeling—will not help you help your organization succeed. In fact, just the opposite is true. Hence, these practices are clearly detrimental to your personal success as an electronic technician (see Figure 9-2).

**How Prejudiced Behavior
Can Harm Your Career**

- Works against teamwork

- Works against peak performance

- Works against continual improvement

- Works against team spirit

- Works against helpful associations

FIGURE 9-2 Prejudiced behavior can undermine your career advancement.

In addition to the obvious philosophical problems, there are some very practical reasons why prejudice and its related practices are bad for your career. These include the following:

- *Detrimental to Teamwork.* Electronics technicians do their work in teams. Hence, being a good team player is critical to your success in the field of electronics. Being a good team player requires that you be cooperative, mutually supportive, and flexible, among other things. Prejudice, stereotyping, discrimination, and labeling work against cooperation, mutual support, and flexibility. Prejudice will undermine the spirit and cohesiveness a team needs to perform at peak levels.

- *Undermines Peak Performance.* One of the most effective ways to help yourself and your organization is to give your absolute best performance at work every day. This is the concept of peak performance, and it is critical to organizations that operate in a competitive environment. It is equally critical to technicians who want to earn promotions, salary increases, bonuses, and the other benefits of outstanding performance. Prejudice, stereotyping, discrimination, and labeling are the enemies of peak performance because they drain you and your teammates of the energy, focus, and motivation needed to perform at peak levels everyday. The truth is, you cannot do your best when distracted by the time and energy-sapping practices associated with prejudice, stereotyping, discrimination, and labeling.

- *Undermines Continual Improvement.* In today's globally competitive environment, circumstances are constantly at work raising the bar concerning what passes for competitive performance. What was considered excellent performance yesterday, may not be tomorrow because your organization's competitors are doing everything in their power to get better and better every day. Consequently, to stay competitive, organizations and the technicians they employ must get better and better at what they do. This is the concept of continual improvement. Your organization's performance will not improve unless your performance and that of your teammates' improves. As with peak performance, prejudice, stereotyping, discrimination, and labeling are the enemies of continual improvement, and for the same reasons. They rob you and your team of the energy, focus, and motivation needed to continually improve your performance.

- *Undermines Team Spirit.* One of the reasons that a cohesive team is able to outperform a group of disparate individuals is the concept of team spirit. Team spirit is an intangible concept that combines pride, mutual-support, identity, loyalty, and cohesiveness in ways that promote peak performance and continual improvement. The "spirit" of a team develops when team members work cooperatively toward the accomplishment of common goals and must depend on each other to achieve those goals. It is difficult to build team spirit and easy to lose it. Few things will undermine team spirit faster than prejudice, stereotyping, discrimination, and labeling. Why would team members who are subjected to these things want to work cooperatively with the very people who act out their prejudices against them? Consequently, discrimination based on irrelevant differences will quickly undermine team spirit.

- *Undermines Helpful Associations.* Successful people do not succeed without the help of others. Even the most talented electronics technicians will need the help of others when trying to build their careers. Remember this: the person you stereotype, label, or discriminate against today, might turn out to be the person whose help you will need tomorrow.

The story of two technicians I will call Bob and Juan illustrates in very practical terms how prejudice, stereotyping, discrimination, and labeling can hurt your career. Bob and Juan grew up in the same neighborhood and went to school together from first grade through high school. They even played together as young children. However, somewhere along the way Bob learned to be prejudiced against Latinos. As a result, throughout their high school years, Bob made a point of labeling Juan and acting out his prejudice in a number of discriminatory ways. By the time Bob and Juan graduated from high school, their relationship was so strained they could not be in the same room together without exchanging harsh words or worse.

After graduating from high school, Juan enrolled at the local community college and studied electronics. Two days after graduation, Juan secured an excellent job in his field. Bob, on the other hand, joined the Navy for a four-year tour of duty. During his time in the Navy, Bob completed several technical courses and qualified as an electronics technician. By the time Bob completed his tour of duty in the Navy, Juan had been working as an electronics technician and had advanced to the level of team leader.

Bob's first goal upon separating from the Navy was to find a good job as an electronics technician. A manufacturing company in his hometown had openings, and Bob applied. His interview went well, and Bob was offered a job. The next day he reported for work. After completing in-processing and a two-hour orientation, Bob was assigned to a team as a service technician. While escorting Bob to his work area, the company's Human Resources Director told him: "You are going to like being part of our service team. You have an excellent team leader—one of our best." Then she said: "Here he is now. Bob, this is your supervisor, Juan."

Overcoming Learned Prejudice

As was stated earlier in this chapter, prejudiced behaviors such as stereotyping, labeling, and discrimination are not acquired at birth. They are learned. The good news in this unfortunate fact is that what can be learned can be unlearned. This section provides strategies you can use to overcome any prejudiced behaviors you might have learned by replacing them with positive behaviors that promote teamwork, peak performance, continual improvement, team spirit, and helpful associations.

Focus on Character Rather Than Race, Gender, Culture, or other Differences

The ways in which people can be outwardly different—race, gender, culture, national origin—are the wrong traits to consider when forming your opinions of people. What we should focus on are those traits that really make people who they are—character traits. Character traits include the following:

- Honesty and integrity
- Selflessness
- Dependability
- Trustworthiness
- Initiative
- Tolerance/sensitivity
- Perseverance

Remember this when you form your opinions of people: race, gender, culture, and the other ways that people can appear outwardly different are not character traits. They do not make people who they are. There are people of every race, both genders, and all cultures who have the kinds of character traits listed previously as well as those who lack them. It is a person's character—not his race, gender, or culture—that makes him who he is.

Look for Common Ground Between People

There is almost always more common ground than differences between people. Even people from different countries who speak different languages, are of different races, and have different cultural backgrounds have more similarities than differences. People of all races, nationalities, and cultures, as well as both genders, tend to share the same desires, ambitions, hopes, fears, and needs.

I can remember going through Marine Corps boot camp at Parris Island, South Carolina in a platoon made up of African-Americans from northern cities, Hispanics from western states, Cajuns from Louisiana, Caucasians from rural southern towns, and a few Asian-Americans from cities on the East coast. Our recruit training platoon was easily the most diverse group of people any of us had ever been in.

At first, all the recruits arranged themselves in groups according to race. This was an attempt to gain a degree of comfort in a distinctly uncomfortable and even alien environment. Having never really spent much time with people of other races, the recruits were simply seeking the comfort of the familiar. This self-segregation into groups by race did not last long though. Our drill instructors made sure of that.

On our first day of training, the entire platoon stood at attention as our three drill instructors demonstrated the Marine Corps' policy on race. One drill instructor was Hispanic, one African-American, and one Caucasian—this fact itself a demonstration of Marine Corps policy. After explaining in no uncertain terms that the only color that mattered from that point forward was Marine Corps green, each drill instructor took out a combat knife and scratched his arm with it—not much, just enough to draw a little blood. Then the drill instructor made a great show of letting us see that the blood of all three was the same color—red. It was a memorable way to let us know that under the skin we were all the same and to make the point that we would be differentiated from that point forward based on character and performance, not race, culture, national origin, or any other feature over which we had no control.

This was a key point our drill instructors emphasized from the first day of training. We had nothing to do with what race, country, or culture we were born into, but we had everything to do with determining our character and our performance. Their point was that in the Marine Corps, we would be judged based on character and performance—things we controlled—rather than race, culture, or national origin—things we did not control. This is a good approach to take at work as you are trying to build a winning career as an electronics technician.

Focus on What Really Matters

Any time a group of people must work together as a team to accomplish goals, it quickly becomes apparent what really matters. When your career and livelihood, as well as that of your team members, depends on getting the job done right and on time day after day, it soon becomes apparent that what matters most about team members is not their race, gender, culture, or politics. Rather, what matters are such characteristics as talent, motivation, attitude, and teamwork; and the undeniable truth is that these characteristics have nothing to do with race, culture, gender, or politics.

My fellow recruits and I learned this lesson many years ago in a memorable way during Marine Corps boot camp. There was a great deal of competition between recruit platoons in such activities as physical training (push-ups, pull-ups, sit-ups, running, climbing ropes, running the obstacle course, and so on), shooting, marching, and classroom studies. People who join the Marine Corps tend to be competitive by nature, and boot camp has a way of bringing this trait out and magnifying it.

For example, a platoon that won a competition on a given day might be rewarded with a little downtime to relax for a few minutes before lights out that night—an incredibly desirable reward in an environment where every second of every day is programmed, controlled, and used to its fullest extent for some aspect of training. A platoon that lost a competition on a given day would be forced to use some of its sleeping time to correct deficiencies. As a result, recruits quickly learned what mattered in the Marine Corps, and it wasn't race, culture, or politics. Rather, it was talent, motivation, attitude, performance, and teamwork.

We had not been in boot camp long before old prejudices, stereotypes, and labels began to be replaced by a new outlook that focused on what really mattered. The recruits we wanted on our teams and in our squads were those who could get the job done so that we could earn the corresponding rewards and recognition rather than extra training conducted when we otherwise could have been sleeping.

Work as an electronics technician is much the same. Rewards and recognition in the form of salary increases, bonuses, promotions, and awards go to those who get the job done. People who help their team get the job done are the *winners* in the workplace, and you will soon find that winning has everything to do with performance, motivation, talent, attitude, and teamwork and nothing to do with race, gender, culture, or politics.

Relate to People as Individuals

The problem with prejudice, stereotyping, and labeling is that if you look hard enough, you can find members of a given racial group, the opposite sex, or other cultures who will live up (or down) to your preconceived notions. There are *some* people from every group that will display the characteristics attributed to that group by prejudiced people. How people become prejudiced is by attributing the characteristics in question to an entire group of people just because they see them in a few members of that group. In other words, the only dependable, logical, and success-oriented way to deal with people is as individuals rather than as members of racial, gender, cultural, or political groups. The minute you attribute a given characteristic to an entire group of people, you are going to be wrong, and when your livelihood and success in your career depend on being right, you cannot afford to be wrong.

In the workplace, there will be people of different races, cultures, and political persuasions as well as both genders who are good team players and are positive, talented, and motivated. Correspondingly, there will be people of different races, cultures, political persuasions, as well as both genders, who will be untalented, unmotivated, and negative. In both cases—good and bad—their work-related traits are the result of their individual character not any group-related factors such as race, gender, culture, or politics.

Review Questions

1. Define the term *diversity* as it relates to your career as an electronics technician.
2. What is meant when it is said that resistance to diversity is learned rather than in-born?
3. Explain the following diversity-related concepts:
   a. Prejudice
   b. Stereotyping
   c. Discrimination
   d. Labeling
4. Describe how acting out prejudice in the workplace can be harmful to your career.
5. List and explain several strategies for overcoming prejudices a person might have learned in growing up.

Discussion Questions

1. In this chapter, an example was given from years ago when it was thought by some NFL and college coaches that African-American athletes would not make good quarterbacks. Based on what we now know about this prejudiced thinking, what would you say to a teammate who made the following statement: "I don't want any women on our team. They don't make good electronics technicians."
2. This chapter makes the case that resistance to diversity is a learned behavior. Discuss how a person might learn to be prejudiced as he or she grows up in America.

3. Do you know someone who has been subjected to prejudice, stereo-typing, labeling, or discrimination? Discuss the situation. What effect did being the recipient of such behavior have on this person?

4. Discuss what is meant by finding *common ground* with people who might appear to be different from you on the outside.

Application Assignments

1. Do some research into the Civil Rights movement of the 1960s in the United States. When were African-American students first allowed to attend previously all-white public schools in the south and what methods were used to keep them out of these schools?

2. Do some research to determine when women were first allowed to vote in the United States. What reasons were given by opponents to defend their position that women should not be allowed to vote?

3. Identify someone who works in electronics, and ask this person if diversity issues come up in the workplace and, if so, what effect they have—both good and bad. Ask for specific examples of anything this person's employer does to promote positive working relationships among diverse groups of personnel.

CHAPTER 10

ADOPT A POSITIVE "CAN-DO" ATTITUDE TOWARD YOUR WORK

LEARNING OBJECTIVES

Upon completion of this chapter, you should be able to do the following:

- *Define the concept of the positive "can-do" attitude.*

- *Explain why a positive "can-do" attitude is important for electronics technicians who are trying to build a winning career.*

- *Explain your responsibilities concerning your attitude on the job.*

- *Describe how your attitude can affect those of your teammates at work.*

- *Demonstrate how to develop a positive "can-do" attitude.*

Winston Churchill was Prime Minister of Great Britain during the difficult years of World War II. He is viewed by historians as the man who almost single-handedly held his country together during the early days of World War II when Great Britain stood virtually alone against Hitler and his Nazi juggernaut. Great Britain was losing the war on every front, and it seemed to many that every day things just got worse. More than anything else, it was Churchill's positive *can-do* attitude that gave the British the strength to hang on and the conviction that eventually the defeats would turn around and they would win. The positive effect of Churchill's can-do attitude can be seen in the comments of his contemporaries:

We owed a good deal in those early days to the courage and inspiration of Winston Churchill who, undaunted by difficulties and losses, set an infectious example to . . . colleagues. . . . His stout attitude did something to hearten his colleagues.[1]

From this quote by one of Churchill's contemporaries, we can see that at a time when his country needed it most, Churchill displayed a positive can-do attitude and used it to help him lead Great Britain to eventual victory in a world war. His is an excellent example for electronics technicians who want to build winning careers in a competitive workplace. The positive can-do attitude Churchill used to help his country eventually win a global war is the same attitude you will need to help your employer win in the global marketplace. Remember, when you help your employer win, you win.

Positive Can-Do Attitude Defined

A positive can-do attitude is an attitude that says whatever the job, I can get it done; whatever the challenge, I can meet it; and whatever the obstacle, I can overcome it. Such an attitude, although certainly optimistic, should not be confused with bragging or making false promises. Rather, a can-do attitude is based on the conviction that, when faced with a challenge, you will do everything in your power that is legal and ethical to meet the challenge, and you won't give up until you succeed in getting the job done. A can-do attitude consists of the following elements (see Figure 10-1):

- Optimism
- Initiative
- Determination
- Responsibility
- Accountability

Elements of a

POSITIVE CAN-DO ATTITUDE

- Optimism
- Initiative
- Determination
- Responsibility
- Accountability

FIGURE 10-1 A positive can-do attitude is not just another form of cheerleading.

I once worked in a team led by a supervisor who had been a technician in the team before becoming the team leader. This supervisor's positive can-do attitude helped our company almost double its gross income in just 18 months. Our company's marketing department had an opportunity to bring in a contract that would be worth almost as much money as all of our contracts for the previous year put together. This sounded good on the surface, but there was a catch. The contract, which normally would have required 3 years to complete, had to be completed in just 18 months. Frankly, most of the people who had to perform the work on the contract had serious doubts that we could complete it in just 18 months. To add to our doubts and our stress, there was a significant penalty for every day the contract was late if we did not complete it on time. In other words, if our company accepted this contract, it had to be completed on time.

During a team meeting called to discuss our part of the work on this contract, most of our technicians expressed doubts, complained that the contract would require more overtime than they wanted to work, and generally grumbled about those "people" in marketing who had no idea what it would take to complete this contract. Our supervisor let everyone vent. Each technician on our team was allowed to get his doubts and concerns off his chest no matter how frustrated or angry he might be in expressing himself.

Then, when everyone had taken his turn at venting, the supervisor calmly smiled and said: "I understand your concerns, but I know you guys and what you can do. In fact, I think you are exactly twice as good as you think you are. We can complete this project in 18 months. In fact, I'm betting that we complete it early." He then went on to tell us that the things people worry about rarely turn out to be as bad as we imagine they will be. His confidence was infectious and, before long, our team—which had the bulk of the responsibility for completing this project—had begun to believe that we could do 3 years of work in just 18 months.

During the 18 months in question, our supervisor was the personification of calm reassurance. He was always positive and upbeat. He didn't let the various obstacles we encountered get him down, and he didn't let them get us down. In addition, he used the crisis presented by this project to convince higher management to let him try some productivity-improvement strategies he had been recommending for more than a year but had been unable to convince management to approve. His strategies worked and became standard operating procedure from that point forward. His positive can-do attitude spread throughout the team, and we eventually did just what he said we would do: we completed the project almost 2 weeks early. Without the calming influence of his positive can-do attitude, we would not have completed the project on time. Higher management knew this and rewarded this supervisor with a promotion that turned out to be one of several until he eventually became a part owner of the company.

Why a Can-Do Attitude Is So Important

Part of succeeding as an electronics technician involves becoming a leader among your teammates at work. You do not even have to be in a leadership position such as supervisor to be a leader among your teammates. You have become a leader when they begin to look to you for a positive example to follow and when you begin to have influence with them because of your positive example.

Because of this, it is important to remember that the attitudes of leaders are often picked up by their followers. In the example given in the previous section, the members of our team of technicians picked up the positive can-do attitude of our supervisor. A leader with a bad attitude is likely to spawn bad attitudes in others. People don't like to follow someone with a bad attitude. On the other hand, as the earlier example of Winston Churchill showed, a leader with a positive, can-do attitude can keep people focused and committed during even the most difficult of times.

"Optimism is also the key to the can-do spirit, to the don't-take-no-for-an-answer attitude that is essential to successful executive leadership. Nearly all-human organizations are subject to an inertia that results in an it-can't-be-done attitude. This was always unacceptable to Churchill."[2] Such an attitude should also be unacceptable to electronics technicians who want to become leaders as part of building a winning career.

Responsibility for Your Attitude

Many people think that a positive can-do attitude is something a person is born with—or without. People often view a positive can-do attitude as a gift people either have or don't have. In reality, nothing could be further from the truth. At the beginning of this chapter, Winston Churchill was used as an example of a leader who was famous for his positive can-do attitude. Churchill is such a good example not because he was born with his famous attitude, but because the reality was just the opposite. Churchill actually suffered from frequent and severe bouts of depression. His famous positive can-do attitude was a hard-won attribute he had to strive constantly to maintain. It is the same with most people.

A can-do attitude can be as much a tool to an electronics technician as an ohm meter or a computer, and just as you are responsible for maintaining the tools of your trade, you are responsible for maintaining your attitude. The following quote is often used to remind people that a positive, can-do attitude is a choice rather than a gift or an accident of birth:

> We cannot choose how many years we will live, but we can choose how much life those years will have. We cannot control the beauty of our face, but we can control the expression on it. We cannot control life's difficult moments, but we can choose to make life less difficult. We cannot control the negative atmosphere of the world, but we can control the atmosphere of our minds. Too often, we try to choose to control things we cannot. Too seldom, we choose to control what we can . . . our attitude.[3]

Your Attitude Affects Other People

John C. Maxwell is a best-selling author of success books and an internationally recognized speaker on the subject. Here is what he has to say about attitudes: "People catch our attitudes just like they catch our colds—by getting close to us. It is important that I possess a great attitude, not only for my own success, but also for the benefit of others."[4]

The fact that people will look to their leaders for the type of attitude they should adopt is precisely why it is so important for you to maintain a positive attitude. In the preceding quote, Maxwell makes the point that an attitude spreads in the same way as a cold.

This is an accurate description of what actually occurs. When one person in a team gets a cold, it seems that before long everybody on the team has one. The same can be said of a bad attitude. Further, just as it is easier to catch a cold than to prevent one, it is easier to spread a bad attitude than to prevent one. This is why it is so important for you to be both persistent and consistent in displaying a positive can-do attitude.

A Positive Can-Do Attitude Is Based on Substance Not Hype

It is important to understand that a positive can-do attitude is not just uninformed cheerleading. In high school, I played on a football team that won only 2 games in a 10-game schedule during my senior year. During the games we lost, our cheerleaders could be heard jumping up and down and yelling about how we were going to come back and win. In several cases, they made these pronouncements when we were more than 40 points behind with less than a minute to go in the game. I can remember being in the huddle as our quarterback called the next play and thinking, "Those cheerleaders must be watching some other game." Their false optimism did not help the morale of our players.

This type of cheerleading is hype and false optimism, not a positive can-do attitude. A positive can-do attitude is based on substance. Optimism is an essential ingredient in a positive can-do attitude but so are initiative, determination, responsibility, and accountability. In the earlier example of the supervisor who assured the technicians in our team that we could complete the contract in 18 months, there was no cheerleading. He knew our abilities as a team better than we knew them ourselves. He knew it would be difficult and that we would have to perform at peak levels. But he also knew that if we took the initiative to effectively implement several productivity-improvement strategies, maintained our determination to perform consistently at peak levels, took responsibility for our performance, and held each other accountable, we could meet the deadline. In other words, his positive can-do attitude was based on substance not hype or false optimism.

Developing a Can-Do Attitude

Although some people are born with a positive can-do attitude, most people are not. Fortunately, you can develop such an attitude. Developing a positive can-do attitude is like developing a muscle: 1) it takes hard work, determination, and persistence; and 2) if you stop working at it, you can quickly lose the gains you have made. This section describes a process for developing a positive can-do attitude.

Step 1: Assess

Do you have a positive can-do attitude? This is an important question that every electronics technician who aspires to a winning career should look in the mirror and ask. The process is called self-assessment, and it can be difficult and even painful. As people, we sometimes find it difficult to confront our personal shortcomings. However, the ability to look at yourself objectively, identify problems, and do what is necessary to correct them is a mark of a mature professional who has the potential to succeed. When assessing your attitude, objectively identify any aspects of your thoughts, feelings, and behavior that need to be improved.

The results of the self-assessment conducted in this step are the beginning point for the plan you will develop in the next step. Use the following questions to conduct the self-assessment. The desired answer to each question is "Yes."

- Are your thoughts about people generally positive?
- Are your thoughts about your work generally positive?
- Can you disagree with people without being disagreeable?
- Can you maintain your composure under pressure?
- Can you perform well under stress?
- In most situations, do you feel like you can get the job done?
- Do you seek responsibility instead of waiting for it to be assigned?
- Do you stay positive when looking for solutions to unexpected problems?
- Does your behavior encourage perseverance when obstacles stand in the way of success?

Step 2: Plan

Your plan should have the following elements: improvement goals based on the self-assessment, specific actions to be taken to achieve the goals, and methods for measuring progress.

Goals

Convert any weakness identified during the self-assessment into a goal. State the goals in behavioral (doing or action) terms as an improvement you would like to make. For example, assume that you answered "No" to the following self-assessment question:

Can you perform well under stress?

This problem area could be converted into a behaviorally stated goal that would read as follows:

Learn to stay calm and focused in stressful conditions.

By converting this weakness into a behaviorally stated goal, you encourage improvement in two ways—both positive. First, you eliminate all ambiguity. Having written down the goal, there is now no question of what you need to improve. Second, you build in personal accountability by creating an expectation—in writing—so that progress can be measured.

Action steps

Identify specific action steps you can take to accomplish each improvement goal you set. Action steps identify more specifically than the goal exactly what must be done to improve. For example, the following action steps would help achieve the goal of learning to stay calm and focused in stressful conditions.

- When I find myself getting anxious in a stressful situation, I will stop, take three deep breaths, and silently tell myself to stay calm.
- When I find myself getting anxious in a stressful situation, I will make a point of focusing on the tasks to be completed rather than my fears in the situation.

Methods for measuring progress

There is a success principle that says, ". . . if you want to make progress, measure it." Measurement is a necessary factor in the formula for accountability. This is why people who are trying to lose weight are required to weigh every day. If your action steps are stated in behavioral terms, they can be measured. This is important because human nature being what it is, you will make better progress if you invest the time and effort to measure results. For example, progress in implementing the action steps from the previous paragraph can be measured. Make note of how many times in a given week you actually stopped and took three deep breaths when confronted with a stressful situation. Make note of how many times you were able to tell yourself in such situations to focus on the tasks to be done rather than your fears relating to those tasks. In each case, is the number more than it was the week before? If so, you are making progress. If not, you must keep working on the problem.

You might wonder why you need to develop a *written* plan for improving your can-do attitude. You might think: "Now that I've completed the self-assessment, I know what needs to be done. I don't need to write it down." However, experience has shown that those who fail to write their plan down also fail to carry it out. Those who invest the time and effort to develop a written plan are more likely to invest the time and effort necessary to carry out the plan.

Step 3: Implement, Monitor, and Adjust

After your plan is complete, implement the plan and use the methods included in it to measure progress. If you are making acceptable progress, continue on course. If not, make adjustments. A plan is just that—a plan. If it's working, stay with it. If not, drop the ineffective strategies, and try new ones. In other words, don't become wedded to your plan. If the plan isn't working, revise it. It is the progress toward improvement that matters, not the specifics of the plan.

Review Questions

1. Define the term *positive can-do attitude* as it relates to electronics technicians who aspire to winning careers.
2. Why is it important for the electronics technician who wants to build a winning career to have a positive can-do attitude?
3. Is a positive can-do attitude something you are born with or something you can develop? Explain.
4. Explain how your attitude can affect the attitudes of others.
5. Explain the steps for developing a can-do attitude.

Discussion Questions

1. Have you ever known someone who has a positive can-do attitude? Discuss with your class the effect this person had on you. Have you ever known someone who had a negative attitude? Discuss with your class the effect this person had on you.
2. Defend or refute the following statement: "Joe is a good technician, but he really has a bad attitude. I don't guess there is anything he can do about the bad attitude though. That's just Joe."
3. Defend or refute the following statement: "I don't believe in all of this attitude stuff. It's nothing but a bunch of cheerleading."
4. Discuss how you would answer the following question from one of your team members: "I want to develop a can-do attitude, but don't know how. Can you help me?"

Application Assignments

1. Do some Internet research or go to your school's library and locate a good story that illustrates the value of a positive attitude at work. Share the story with your class.

2. Conduct a self-assessment of your attitude, and develop a plan for improving any weaknesses you identify.

3. Talk to someone you know who works in the field of electronics or in some other field if you don't know someone in your field. Ask this person to give you his or her views on the effects of positive and negative attitudes at work. Share what you learn with your class.

Endnotes

1. Maurice Hankey as quoted in Steven F. Hayward, *Churchill on Leadership*, (Forum, Rocklin, CA: 1998), 115.
2. Steven F. Hayward, 116.
3. Anonymous, "Attitude," *Barlett's Familiar Quotations*, ed. Emily Morison Beck, (Little Brown, Boston: 1980), 413.
4. John D. Maxwell, *Developing The Leader Within You*, (Thomas Nelson Publishers, Nashville: 1993), 105.

CHAPTER 11

LEARN HOW TO RESOLVE CONFLICT ON THE JOB

LEARNING OBJECTIVES

Upon completion of this chapter, you should be able to do the following:

- *Demonstrate how to handle complaints from teammates and customers.*
- *Demonstrate how to resolve conflict between other team members.*
- *Explain how to deal with angry people at work.*

The most successful electronics technicians are good at resolving conflict. This, in turn, makes them good at dealing with difficult people, whether those people are teammates or customers. This is an important *success skill*. Today's competitive and often stressful workplace can be a virtual factory for conflict. People in organizations have different agendas, ambitions, opinions, backgrounds, and perspectives. These differences as well as all of the other ways that people at work might be different can contain the seeds of conflict. Add two other ingredients—ego and self-interest—and you have a potentially volatile mix, especially when people work together under the pressure imposed by competition, deadlines, ambition, and the profit motive.

Many people have developed the unfortunate tendency to see others who differ with them as being not just wrong but bad. Consequently, conflict in organizations is a common occurrence. If this counter-productive situation is allowed to get out of hand, an organization can get so bogged down in negativity that it cannot perform at the level needed to be competitive. This is why learning how to resolve conflict is one of the keys to advancing your career as an electronics technician.

To achieve consistent peak performance and to continually improve, organizations need their personnel at all levels and in all positions working cooperatively toward achievement of the same goals. This does not mean they always have to agree—quite the contrary. Differences of opinion about how best to do the job can lead to better decisions, better solutions, and better performance if handled professionally (that is, if those involved can disagree without becoming disagreeable). Your opinion about the best way to do something might challenge the thinking and sharpen the opinions of others in a positive way provided that everyone involved is able to discuss their differing opinions in an intelligent, mature, and professional manner without anger, hurt feelings, and grudges.

Performance cannot be continually improved in a counter-productive environment where conflict has been allowed to turn negative and become personal. In fact, in such situations, performance typically declines. To find ways to continually improve, people need a work environment that supports the free flow of ideas, and they must be able to disagree without being disagreeable.

The most successful electronics technicians have learned how to prevent and resolve conflict on the job. This is one of the reasons they are successful. Electronics technicians who are good at conflict prevention and resolution know how to keep disagreements on the high road that leads to better ideas as well as how to pull it up from the low road that leads to anger, hurt feelings, and grudges.

Handling Complaints from Teammates and Customers

As soon as you get your first job as an electronics technician, you will begin to hear complaints. All people in the workplace—from the newest technician to the CEO of the organization—hear complaints from teammates and customers. Complaints that are handled well can prevent minor problems from blowing up and becoming major problems, defuse conflicts before they get out of hand, lead to changes that result in improvements, and turn unhappy customers into satisfied customers.

On the other hand, complaints that are ignored or handled poorly can lead to disaster. For example, many of the episodes of workplace violence that occur every year begin as minor complaints that fester and grow over time because they are ignored by people who could have done something about them. Complaints that come from inside an organization that are ignored or handled poorly can lead to quality, productivity, and morale problems. Complaints that come from outside the organization that are ignored or handled poorly can lead to problems with customers, suppliers, and regulatory agencies.

Once you become a member of a work team, no matter how busy you may be, when a teammate or a customer approaches you with a complaint—*listen*. If the complaint falls outside of your area of responsibility and authority, put the complainer in touch with the person who can help him. A good rule of thumb is to give people who complain to you five minutes of uninterrupted time to vent—to get the problem off their chest. After this person has made his or her complaint, ask what the person recommends as a solution.

If the person making the complaint happens to be from outside of the organization—particularly if he or she is a customer—it is a good idea to say "thank you." After all, instead of complaining, this person could have just taken his or her business to one of your competitors. The first step in handling complaints is always the same—*listen*.

Listening Skills

The most important skill for handling complaints, resolving conflict, and dealing with difficult people is listening. If you can be a good, patient listener, you can be effective when it comes to handling all of these types of situations. When handling complaints, whether they come from teammates or customers, try the following strategies to improve your listening skills (see Figure 11-1):

- *View the complaint as a potential opportunity to make improvements.* As an electronics technician, you will be a busy person. Most of your days on

---

Listening Strategies
for Handling Complaints

- Look for opportunities in complaints.

- Eliminate distractions.

- Look the complainer in the eyes and look interested.

- Apply common sense.

- Watch for nonverbal cues.

- Paraphrase and repeat the complaint.

- Maintain your composure.

FIGURE 11-1 These strategies will help when handling complaints.

the job will be full even without the added responsibility of listening to people who want to complain. Consequently, there will always be the temptation to treat complaints as unwelcome intrusions that you do not have time for. Avoid this temptation no matter how busy you are with your normal duties. Electronics technicians who ignore complaints because they are too busy to deal with them are really saying, "I'm too busy to make improvements, retain a customer, prevent problems, or diffuse conflicts." Electronics technicians who are too busy to do these things will be lucky to keep the positions they have, much less advance in their careers.

- *Eliminate distractions that interfere with listening.* When someone approaches you to make a complaint, eliminate distractions that might interrupt the conversation or distract your attention from what is being said. People with complaints are already frustrated. You don't want to add to their frustration by allowing telephone calls or other distractions to interrupt their dialogue with you. Give people who want to make a complaint your undivided attention for five minutes. Even serious complaints seldom take longer than this.

- *Maintain eye contact and an expression of interest.* I once worked with a person who had an interesting way of "listening" when people brought complaints to him. He would sigh or groan loudly, put his head in his hands, and ask, "What is it this time?" Clearly, this person did not want to hear complaints from his teammates, direct reports, or customers. His "shoot-the-messenger" attitude eventually backfired on him. Because he did not want to hear what his team members had to say, they stopped telling him anything—even things he really needed to know. The predictable result was that his ignorance of problems he should have known about eventually caused even bigger problems; a fact that hurt his career.

- *Listen with common sense.* This means to listen between the lines. Sometimes what the complainer doesn't say is just as important as what he does say. If you sense that the complainer is leaving out pertinent facts, exaggerating, or bending the facts in his favor, he probably is. In other words, if common sense tells you that something about the complaint sounds wrong, you are probably right—something is wrong. Common sense will tell you when something you hear in a complaint is not right if you will listen to your common sense. If your common sense tells you that pertinent facts are missing from the complaints, wait until the person has said what she has to say and then ask: "There are a few things I don't understand about your complaint. How about helping me fill in some missing information?"

- *Look while you listen.* You can learn a great deal from a conversation by looking while you listen (that is, watching for nonverbal cues). Even without realizing it, people send messages with their eyes, tone of voice, facial expressions, gestures, and posture. If the complainer cannot look you in the eye or maintain steady eye contact, something is wrong. He might be exaggerating or even lying to you. Even if he is just holding back information, the fact is for some reason he is too uncomfortable with what he is saying to maintain eye contact. If the complainer's voice becomes strained and quivers, she is probably nervous. Whether you realize it or not, you know how to read nonverbal cues. In fact, you could understand nonverbal communication before you could understand spoken or written communication. A child can tell if his mother is happy, sad, nervous, stressed, or angry even before he can understand the first spoken word. The child can tell because his mother gives off nonverbal cues (for example, how tense she is as she holds the child and her tone of voice as she talks). The key to understanding nonverbal communication is not to look for specific gestures. Rather, it is to watch for agreement or disagreement between what is said verbally and what is "said" nonverbally. When you notice disagreement between the verbal and the nonverbal, ask the complainer for clarification. Inconsistencies in what the complainer says should be cleared up right away. You do not want to invest time and effort in solving the wrong problem.

- *Paraphrase the complaint, and repeat it back to the person who made it.* It is important for the person who is complaining to know you have listened, and it is important for you to accurately understand the complaint. The best way to make sure this happens is to paraphrase the complaint and repeat it back to the complainer. Paraphrasing means putting what was said in your own words. For example, assume a teammate complains that she is expected to work more overtime than any one else on this team. She is clearly upset and, as a result, makes several negative comments about her teammates and the company. After she has said what she has to say, you feel sure you understand the problem. This is the point where you paraphrase the complaint and repeat it back to her. In this case, you might say, "Let me make sure I understand what you are saying. You are upset because you've been asked to work overtime twice this month, and it is interfering with your night classes." By listening to what was said as well as what wasn't said, you were able to get to the heart of the complaint and summarize in a very brief statement. Her comments about being asked to work more than anyone else were obvious exaggerations because everyone on the team had been asked to work overtime twice this month.

- *Stay calm.* People who make complaints are often angry and emotional. The natural human tendency in such a case is to respond in kind, that is, to answer anger with anger. Returning the anger of a person who makes a complaint is the wrong approach. When listening to complaints from customers, teammates, or any one else in the workplace, it is important to maintain your composure; stay calm. Maintain a non-judgmental, neutral attitude. Responding to anger with anger of your own only makes matters worse. But if you stay calm and let the complainer vent, his anger will begin to dissipate, and you will be able to begin discussing solutions. This is the main reason for staying calm. By staying calm, you can help the other person settle down and make the transition from anger mode to solution mode.

Resolving Conflicts Between Other People

After you achieve the position of team leader or supervisor, you will need to be effective at helping to resolve conflict between other people. When counter-productive conflict occurs between team members, it can quickly spread as other team members choose sides in the battle. Conse-

---

Conflict Resolution Strategies

- Help participants identify the source of the conflict.

- Encourage participants to look for solutions rather than problems.

- Encourage participants to take responsibility for resolving the conflict.

- Guide participants toward an appropriate resolution.

FIGURE 11-2 These strategies will help when team members are engaged in conflict.

quently, it is important to get counter-productive conflict resolved quickly (see Figure 11-2).

When you have team members involved in conflict, don't wait to see what might happen—take action right away. Bring them together in a private setting; give them an opportunity to state their grievances without interruption, contradiction, or judgmental comments from you or each other. Treat the team members with respect and let them know you take their conflict seriously. The message they need to hear is this: "You are both important to the team, and the team needs you to work together co-operatively for the good of all of us. Let's talk about how that can happen."

After each party to the conflict has stated his or her case without interruption or contradiction, let them both know that the goal is resolution and cooperation. The following strategies can be used to move them toward resolution and cooperation: 1) identify the source of the conflict (Is it about poor communication, insufficient resources, differing perspectives, differing values, differing agendas, and so on?), 2) remind participants that all issues are to be discussed in a mature and positive manner, 3) remind participants that they are responsible for resolving the conflict, 4) ask participants to propose solutions and then discuss what they propose, and 5) guide participants toward a positive resolution of the conflict (that is, a resolution that serves the best interests of the organization).

Identify the Source of the Conflict

People in conflict tend to turn the other person into the "bad guy." However, just disagreeing with someone does not a make a person bad or even wrong. In reality, there are several factors that can be the cause of workplace conflicts. These factors include poor communication, differing perspectives, differing values, insufficient resources, and differing agendas to name just a few. It is important to your teammates to see that their conflict is probably based on logical, understandable factors rather than on malice, bad intentions, revenge, jealousy, power, greed, ego, ignorance, and other negative issues that make the other person a "bad guy."

Communication problems play a part in almost all incidences of workplace conflict. Even when poor communication is not the root cause of the problem, it is often a contributing cause. Having listened to both participants state their positions, does it appear that poor communication is contributing to the conflict? If so, let both parties know because a disagreement that results from poor communication is an easier cause of conflict to accept than the "bad guy" causes people tend to attribute to others who disagree with them (for example, malice, greed, power, ego, ignorance, and so no). If poor communication appears to be a contributing factor, point this fact out to both parties.

People can have widely differing points of view, values, and perspectives concerning the same issue. For example, think of a candidate in a presidential election. Some people will think he is wonderful, whereas others will think he's the worst person who ever ran for office. How can people look at the same individual and see him so differently? It's simple—they have different perspectives, values, and points of view. Some people are liberal, and others are conservative. Some people are extroverts, and others are introverts. Some people are big-picture oriented, and others see only the details. When someone has a different perspective than you, the tendency is to ascribe the difference to something bad or wrong in the individual (the "bad guy" syndrome introduced earlier). You can help resolve conflict by helping your teammates who are in conflict realize that having a different perspective or different values does not make another person bad or even wrong. Sometimes the issue is not who is right or wrong, but who has the better idea in terms of the organization's mission and goals.

Promote a Solution-Oriented Discussion of the Conflict

At this point in the process, I remind each party in the conflict of the organization's mission statement. I explain that our job is to help the organization accomplish its mission. Personal agendas have no place in the discussion. I explain that the two parties have been brought together to have a mature and positive discussion about issues, and that everything said during the discussion should be viewed in the context of how it helps the organization accomplish its mission.

This is just another way of saying, "It's not about you—it's about us." Often, when people in conflict get this message, the conflict goes away, and nothing further need be done. However, if the discussion does need to continue, you have at least made the transition from petty bickering to a mature discussion that might lead to a resolution.

Remind Participants That They
Are Responsible for Resolving the Conflict

When people in conflict bring their disagreements to you, they are sometimes trying to do what I call a *hand-off*. In other words, they want to hand off their problem to you and let you solve it for them. This is like two boxers turning their fight over to the referee and asking him to finish it. Do not let this happen to you. Your role is more that of the mediator than the judge when trying to help teammates resolve conflict. Let participants know that they are responsible for resolving the conflict. Your role is to facilitate the process—to help *them* solve *their* problem.

Guide Participants Toward a Solution

The best solutions to conflicts are those that are worked out by the participants themselves. Consequently, after you have helped your teammates identify the real source of their conflict, reminded them that a mature, positive discussion is the best approach, and explained that they are responsible for resolving the conflict, ask each of them to propose a solution. Listen to what each person says, and don't allow interruptions, comments, or judgmental behavior from either person when the other is talking. Your goal in this step is solicit proposed solutions that can then be discussed in terms of what is best for the organization.

In this step, you are trying to guide participants toward the optimum solution—the solution that comes closest to helping the organization achieve its goals. An effective way to do this is to use the organization's mission statement and goals as the basis for conducting a quick cost-benefit analysis of each solution proposed. Don't make the mistake of trying to work out a compromise solution that will please both parties and, as a result, make you popular. This is not politics you are engaged in—it's business. The most popular solution is not necessarily the best solution from a business perspective.

Dealing with Angry People at Work

Learning how to deal effectively with angry people will help your career. Whether the angry person is a fellow team member, a customer, or anyone else at work, you can increase your value to your organization by learning to do three things: 1) stay calm when confronted by an angry person, 2) apply proven strategies to calm down the angry person, and 3) help angry people make the transition from anger mode to solution mode.

How to Stay Calm When Confronted by an Angry Person

When dealing with angry people, the natural tendency of people is to do one of two things, both of which are wrong: 1) respond in kind with their own anger, or 2) shrink away because of the other person's anger. Although both reactions are understandable, they represent the two worst responses you can employ. The most successful electronics technicians have learned the cardinal rule of dealing with angry people: *When faced with an angry person, if you lose your temper, you lose period, and if you shrink away you lose for good.* When you respond to an angry person with anger of your own, the situation is probably just going to escalate and get worse. On the other hand, when you shrink away from the other person's anger, you are likely to embolden him. The best approach is to stay calm, stand your ground, and maintain a nonjudgmental attitude.

Of course, staying calm when the other person is angry can be difficult. Try the following strategies for staying calm when confronted with anger. First, take a few deep breaths to settle your breathing and keep your pulse rate normal. Second, ignore the anger and listen past it for the substance of the problem—try to determine what is really bothering the angry person. Angry people often become inarticulate and unable to express themselves in anything but crude terms. To further complicate matters, they tend to exaggerate, leave out important information, add irrelevant facts, and generally make more noise than sense.

The best way to stay calm in such situations is to simply ignore the *noise* of the angry person and focus instead on listening past it for any facts that might help you understand what is really bothering the angry person.

When dealing with an angry person, if you get hung up focusing on the *noise* of the anger, you might miss important cues for identifying the actual source of the problem. Take a couple of deep breaths, don't shrink away, ignore the anger, and listen for cues that will lead you to the source of the problem. The harsh words, exaggerations, and even threats associated with anger are symptoms of the problem, not causes. Do not allow yourself to be distracted by them.

There is one caveat that must be made at this point. We now live in a world in which some people act out their anger in violent ways. Consequently, when dealing with an angry person, it is important to apply common sense and intuition. Typically, anger in the workplace is just a volatile emotion that will settle down after the angry person gets it off his chest. However, when dealing with an angry person, do not take chances, especially if you don't know the individual in question. If your intuition and common sense tell you that the person is on the verge of violence, detach yourself from the situation immediately and seek help.

Calm the Angry Person

If you can manage to stay calm when confronted with anger, you may be able to help the angry person calm down. Following are some strategies you can use to help an angry person get a grip on his or her emotions:

- *Listen in a nonjudgmental manner while letting the angry person vent without interruption.* Angry people are like steam-powered boilers. They must be able to vent or they might explode. Consequently, one of the best strategies for helping an angry person calm down is to just let him vent. Let the angry person get it all off his chest while you listen without interrupting or contradicting—even if what he is saying is obviously wrong. While listening, look directly at the angry person with a nonjudgmental look on your face. Don't give off any negative nonverbal cues such as shaking your head in disagreement, shrinking away defensively, crossing your arms, or rolling your eyes. Just listen and let the angry person vent.

- *Acknowledge the anger.* Acknowledging another person's anger is usually an effective way to begin calming that person down. You can do this by saying something as simple as: "I can see you are really angry about this." Most people, having been given an opportunity to vent without interruption or contradiction and then having had their anger acknowledged, will begin to calm down. If the angry person you are dealing with does not calm down, continue to listen, continue to let her vent, and acknowledge the anger again later in the conversation.

- *Use an apology as a bridge to get from the anger mode to the solution mode.* You are trying to help the angry person make the transition from the anger mode to the solution mode. In other words, you are trying to help them stop focusing on why they are angry and start focusing on how to solve the problem in question. Sometimes a brief and simple apology will serve this purpose. An example of a brief and simple apology is as follows: "I am sorry you are angry. Let's see what we can do to correct the problem." Notice that you are apologizing because the person is angry, not for any other reason. You have not said that you or anyone else did anything wrong—even if this is the case. Instead, you have apologized because the situation in question made the person angry and then you immediately made the transition to the solution mode (that is, "Let's see what we can do to correct the problem.").

- *Paraphrase and repeat back.* After you have made the transition from the anger mode to the solution mode, paraphrase the person's problem as you understand it and repeat it back in your own words. This will let the person in question know you have listened or give him an opportunity to correct your perception if you have misunderstood what he said. This is done to ensure that if there is a problem you need to solve, that you get focused on the right problem.

- *Ask open-ended questions to clarify and to solicit additional information if necessary.* If the person in question proposes a solution or makes a point that is not clear or that does not make sense, use open-ended questions to clarify and to solicit additional information. An open-ended question cannot be answered "yes" or "no" or with just one word. You can turn any question you have into an open-ended question by beginning it with such phrases as: "Tell me about...," "What do you think about. . .," or "What are your thoughts on. . . ." Open-ended questions are the best way to get information from a person.

- *Confirm any solution agreed to.* Angry people sometimes just want to vent—to get their feelings off their chest while someone listens. When this is the case, you have solved the problem by just listening. However, sometimes an angry person will want you to do something, and you will agree that something needs to be done. When this is the case, make sure that the person in question understands the proposed solutions and all of its ramifications. Never assume that the other person understands what you plan to do to solve the problem without first confirming the solution and all of its ramifications. A good solution is one that, once put in place, will stay in place. If the person in question does not understand some aspect of the solution proposed, she might just become angry again when things do not work out the way she thought they would.

Review Questions

1. What are the five main things you need to know how to do to resolve conflict and deal with difficult people?
2. Explain how you should handle complaints made by employees or customers.
3. Describe how you can use complaints to make improvements.
4. Explain the process for helping other people resolve conflicts they are having with each other.
5. Explain how you should handle the situation when the work of employees who report to you is being negatively affected by their personal problems.
6. Explain the various strategies for dealing with angry people at work.

Discussion Questions

1. Discuss how you would handle the following situation: An electronics technician on your team (assume you are the team leader) stops you in the hallway and complains that she can't get her work done on time because the other members of the team spend too much time goofing off instead of working.
2. Discuss how you would handle the following situation: You have two team members who just do not seem to be able to get along, and their constant picking at each other is beginning to affect the work of other team members. In fact, the performance of the team is going downhill as the result of the fights between these two technicians. Both are good electronics technicians that you need on your team, but you also need them to get along.
3. Discuss how you would handle the following situation: One of your team members, Janice, is having personal problems at home. Her personal problems are beginning to affect the quality and quantity of her work. In addition, her fellow technicians are beginning to complain about her.
4. Discuss how you would handle the following situation: One of your technicians interrupts a team meeting and says: "I am sick of all this

talk about customer complaints. Why doesn't our boss come with us on some of these service calls and see if he can do any better?" Then he gets up and stomps out of the meeting slamming the door as he leaves.

Application Assignments

1. Do some research into a case of workplace violence. Find one of those cases in which an employee of an organization becomes angry for some reason and acts out the anger in a violent way in the workplace. Analyze the case to determine if there might have been ways the incident could have been prevented.

2. Talk to a supervisor or team leader in a local company. Ask this person how he or she handles conflict between direct reports.

3. Identify someone who works in electronics, and ask this person to explain how conflict on the job affects productivity and morale. Ask for specific examples that you can share with your classmates.

CHAPTER 12

PERSEVERE WHEN THE JOB BECOMES DIFFICULT

LEARNING OBJECTIVES

Upon completion of this chapter, you should be able to do the following:

- *Explain the importance of perseverance in building a winning career as an electronics technician.*

- *List and explain several perseverance strategies for electronics technicians.*

- *Explain how electronics technicians can face and overcome adversity.*

- *List and explain several strategies that will help you keep going when the job becomes difficult.*

A career is like a baseball season—it consists of more than one game. In a baseball season, there are many games. Even the best teams will lose some games, but they are the best because they don't give up and don't quit. The best teams persevere. They keep giving it their best even when losing games, and in this way, eventually emerge as winners. Successful electronics technicians are like this. They are successful because they refuse to let occasional failures cause them to become frustrated and give up. You are going to make mistakes. You are going to attempt something and fail. But one of the main characteristics of successful people in any field is that they turn their setbacks in life into comebacks. In other words, they persevere, learn from their mistakes, and keep trying.

Perseverance will be an important characteristic to develop as you work to build a winning career in electronics. There will be times when things won't go your way. You might not get the recognition you deserve for doing a good job. You might not get that promotion you hoped for or that raise you really needed. The sad truth is that life is not always fair, neither in general nor in the workplace. It is at times like these—the inevitable times when life gets hard and the job becomes difficult—that perseverance is so important. One of the most important lessons you can learn before beginning your career in electronics is that you are going to have setbacks; count on it.

Things that should go your way sometimes won't, and the reasons they don't might seem unfair to you. When this happens, don't waste time and energy focusing on the unfairness of it all and don't get down on yourself if you made mistakes that contributed to the failure. Instead, look objectively at any mistakes you might have made, and ask yourself this question: "What could I have done differently that might have made this situation turn out better." If you made mistakes that contributed to failure in the situation in question, learn from them, and don't make the same mistakes again. If you made no mistakes, and things still turned out badly, chalk it up to the fact that life can be unfair sometimes, and then move on. Many people fail over and over because they get caught up in focusing on the unfairness of the situation when what they should be doing is learning from it, moving on, and persevering until they finally succeed. The most successful people in any field are the ones who learn to view failure as a temporary setback that can and will be rectified.

It has been said that the great inventor, Thomas Edison, failed more than 10,000 times in his attempt to invent the incandescent light bulb. What is important in the case of Edison is not that he failed so many times, but that he never gave up and, as a result, eventually succeeded. The next time you enter a dark room and flip on the light switch, think about the perseverance of Thomas Edison. Think of him when as an electronics technician you attempt some assignment and fail. When this happens, follow Edison's example. Scratch the approach that failed off of your list, and try again—this time using another approach. Think of

Edison any time things don't go well in your career, and you find yourself getting discouraged. Those who persevere typically win in the long run.

Don't Quit and Don't Give Up

Building a winning career as an electronics technician will require more than knowledge, skills, and education. The world is full of knowledgeable, skilled, educated people who either achieve only mediocre careers or fail altogether. Often when a knowledgeable, skilled, and educated person fails, the culprit is a lack of perseverance. Some people, in spite of excellent qualifications, have no gumption. When faced with tough times, they get frustrated and simply give up. This is a tragedy because, so often in life and at work, the person who ultimately wins is the one who is willing to keep going a little longer than the competition.

I can remember playing football during my service in the Marine Corps. It was the last game of the season—one of those low-scoring, tough-as-nails defensive battles. Our offense had been unable to move the ball against our opponent's defense. Fortunately, our defense had been equally tough on our opponent's offense. The score was tied at six points a piece and we had the ball at mid-field. I played tight end. It was a hot, humid day and all of the players on both teams were worn out. We could hardly stand up much less run the ball, block effectively, or pass.

Our coach called a timeout, and we all huddled on the sideline. In a calm, almost matter-of-fact voice he said: "Right now is when the real game begins. Every one of you are tired, and so is every player on the other team. Everything is even at this point. The team that decides to persevere and give everything its got for one more series of downs is going to win this game. I think you are that team. Get out there and give it everything you've got. When you get to the point where you think you cannot take another step, take that step. That's how we will beat these guys."

The coach was right. We went back on the field, lined up, and drove the ball yard-by-yard toward the goal line. It wasn't fancy, and it wasn't pretty, but we just kept going no matter how badly we wanted to stop. Finally, the other team gave up. We could feel it when it happened, and shortly thereafter, we put the ball in the end zone. We weren't a better team than our opponent. In fact, I don't think I ever played in a game in which the two teams were so evenly matched. When we finally won, it was because we persevered when our opponent didn't—not because we had more talent. Throughout my career—more than 40 years so far— I have seen people succeed simply because they were willing to persevere, and I have seen people fail simply because they weren't.

A Case Study in Perseverance

My favorite story of perseverance is about a man I am proud to call my friend. He is Colonel George "Bud" Day, an Air Force pilot and prisoner of war during the Vietnam conflict who was awarded the Medal of Honor for his courageous perseverance in the face of incredible adversity.

In 1967, Day was a Major in charge of a squadron of F-100 jets nicknamed the "Misty Squadron." The pilots of the Misty Squadron were forward air controllers who flew missions over Communist territory in North Vietnam in an attempt to spot targets on the ground for American bombers to attack. On one of these missions, Day's jet was hit by ground fire from the enemy sending it into a steep dive. He had only enough time to bail out before the plane crashed in a ball of flames in the jungles of North Vietnam. As Day ejected from the damaged jet, he was thrown against the fuselage breaking his arm in three places. But his parachute opened in time, and he floated down into the waiting arms of enemy soldiers. Upon landing, Day twisted his knee; adding to his injuries and making it difficult to walk.

In spite of his painful injuries, the enemy soldiers forced Day to walk to an underground shelter where they could interrogate him. The interrogation was vicious and brutal. It included a mock execution and hanging an already injured Day upside down by his feet. Yet in spite of his painful injuries and the brutal torture, Day persevered—he steadfastly refused to cooperate with his interrogators who wanted information that would help them shoot down additional pilots; especially any who might come looking for him.

The torture continued for five days until Day appeared to be so beaten up physically that he could not possibly escape. As a result, on the fifth day of captivity the guards failed to watch Day as closely as they should have, and he quietly slipped away into the jungle. It was hours before anyone noticed that the prisoner was missing. Day had used this time to put as much distance as his wounded and beaten body would allow between himself and his captors.

When he had gained a good head start on his captors, Day began to sleep as best he could during the daylight hours and travel mostly at night. After only two days on the run, while sleeping in a jungle thicket, Day was startled awake by a huge explosion. An American bomb or missile had exploded nearby driving hot shards of burning shrapnel deep into his leg. Because he was asleep when it was dropped, Day never knew what the bomb that hit him was aimed at, but it's likely that an American pilot spotted one of the Viet Cong patrols that was searching for Day and, unable to see him hiding there in the jungle, aimed his ordinance at the enemy patrol.

Day's inventory of painful injures included a broken arm, twisted knee, shrapnel wounds, and cuts and bruises from the torture that, without treatment, were quickly becoming infected. In spite of his mounting injuries, Day continued to slowly and painfully hobble in a southerly direction as fast as he could manage. His goal was to reach friendly territory in South Vietnam, and rejoin his squadron. But his wounds, the jungle environment, and a lack of food and water began to take their toll.

After approximately 10 days with no food other than an occasional frog and some wild berries, Day was suffering from dehydration, starvation, and a host of wounds and injuries. In spite of this, he persevered. With every ounce of strength he could muster, the determined, courageous pilot continued his trek toward freedom and safety. Then, one day as he limped along through the jungle, Day heard the noise of helicopter rotors. Knowing they would be American choppers, he summoned all of his remaining strength and hobbled toward the noise.

Up ahead in a clearing, Day could see American helicopters evacuating a U.S. Marine Corps unit from the bush. If he could just manage to get to the choppers before they took off, he would be home free. Day knew those choppers represented his salvation, so he put everything he had left into covering the ground that separated him from them. Unfortunately, slowed by his catalog of injuries, he arrived too late. In a heart-breaking twist of fate, just as he hobbled into the landing zone the choppers took off never having seen or heard him.

Day could only stand by and watch helplessly as the helicopter that could have been his salvation lifted into the sky without him and flew away. Suddenly Day's thoughts were wrenched violently back to the reality of his predicament by the sound of enemy gun fire. The Viet Cong patrol that had been searching for Day, had finally caught up with him. As he limped painfully toward the beckoning cover of the jungle, his Viet Cong pursuers finally found their mark. Day was shot in the hand and the leg. Shortly thereafter, in spite of persevering for almost two weeks against the worse possible odds and in the worse possible conditions, this courageous Air Force pilot was recaptured.

Day was returned to the same underground shelter from which he had escaped, but this time, the enemy soldiers had a better idea of the type of man they were dealing with. Rather than try to gain his cooperation by torturing him again, they shipped Day to the infamous prisoner of war camp in Hanoi known to Americans euphemistically as the "Hanoi Hilton."

When Day arrived at the Hanoi Hilton, he was suffering from malnutrition, infected wounds, exhaustion, and loss of blood. In response to the pain he endured from torture and an assortment of wounds, Day's hands were curled tightly into claws that he could not open. When he was roughly thrown onto the cold and damp concrete floor of his jail cell, Day could neither feed nor dress himself. Thus began a period of constant torture, humiliation, starvation, and degradation that would last more than five years; a period during which Day's determination to persevere would be tested daily in ways that most people cannot even imagine.

Having thrown Day into a cell, his captors simply left him—thinking they were leaving him to die. But again they underestimated the determination and perseverance of this courageous American. With the help of his cell mates, Day was eventually able to regain the use of his hands and mend his broken bones and wounds sufficiently to function on a daily basis.

As one of the senior officers in his cell block, Day knew it was his responsibility to set an example for the others. He determined that he would never give in to the North Vietnamese Communists, never quit on his men or his country, and endure whatever he had to in order to do his duty as an officer and maintain his dignity as a human being. This admirably approach to a nightmarish situation would cost him dearly.

During his more than five years as a prisoner of war, Bud Day paid a heavy price for his courage and perseverance. His captors knew that in order to break the other American prisoners, they would have to break the officers who they looked to as role models, and for many of the American prisoners of war, Bud Day was one of these officers. Consequently, Day was systematically and brutally tortured on a regular basis. He was starved, beaten, threatened, and humiliated, but through it all, he continued to set an example of persevering in the face of extraordinary adversity.

To give the men something to help them endure their ongoing ordeal, Day and other senior officers began conducting religious services, an act strictly forbidden in a Communist prison. One day while conducting a service, North Vietnamese guards burst into their cell with rifles locked, loaded, and aimed right at them. Day knew that the guards would have no qualms about shooting American prisoners, but he also knew that leadership demanded a strong stand at this point. Consequently, he stood up, faced the guards who were now pointing their weapons at him, and began singing America's national anthem. Soon the other prisoners, moved by the courage of his example, joined in and made it a chorus. The Communist guards were so stunned by this simple act of courageous leadership that they simply shouldered their rifles and backed out of the cell.

Day was finally released from captivity on March 14, 1973, ending an almost five-year ordeal that few people in the world will ever experience or could possibly endure. But through it all, Day persevered, refused to quit, and never gave up. As a result, on March 6, 1976, Bud Day was awarded our nation's highest decoration for valor in combat—the Medal of Honor.

There are going to be times when your duties as an electronics technician will seem overwhelming, when the job will become so difficult you will feel like giving up and quitting. When you face tough times such as these, remember the example of Colonel Bud Day, and just keep going.

What to Do When You Want to Quit

Have you ever had to exercise (run, do push-ups, sit-ups, pull-ups, and so on) until you felt like you could not possibly do anymore? Have you ever had to stay up late studying for a test until you felt like you could not possibly stay awake another minute? What you needed in situations such as these is perseverance. Perseverance is running a little farther when you feel like you cannot take another step. It's doing another push-up when your arms ache, your muscles burn, and you want more than anything to just quit. It's studying for another 20 minutes when you feel like you are ready to fall asleep.

Although these are physical examples, perseverance is more mental than physical. To persevere, you have to consciously and mentally will yourself to go on when that's the last thing you want to do. It's about mentally telling yourself to work a little longer, try one more time, and just hang in there a little longer when you want to give up. At times in your career, this is exactly how you will feel. You will be tired, frustrated, and exasperated, and you will want to just give up and quit.

There will be times when you will feel like saying: "I will never get this device to work," "I will never please this boss," "I will never get that raise I need," or "I will never get the promotion I want." When you feel this way—like you want to quit—it is especially important to persevere and to just keep going. The reason for this is simple. After you give in to adversity, giving in can quickly become a habit, and when this happens, your chances of building a winning career are gone.

The following strategies are provided to help you when difficulties in your career make you want to quit. When you feel that way—and believe me there are going to be plenty of times when you will—try the following strategies. They might help you keep going a little longer instead of quitting.

- When you feel like giving up and quitting, think of the great inventor, Thomas Edison. In trying to invent such useful products as the light bulb and the storage battery, he failed repeatedly. Legend has it that it took him almost 10,000 attempts to finally succeed in finding the right filament for the incandescent light bulb and the right materials for the storage battery. Although frustrated, Edison never quit trying. He simply refused to give up, and as a result, the great man finally succeeded.

- Before you get to the point where you feel like quitting, you have already put in a lot of time and effort on the task in question. Think about that when you get frustrated and feel like giving up. Think about what the great football coach from the old days, Vince Lombardi, used to tell his players: Paraphrased, Lombardi's message was: *The harder you work at accomplishing a goal, the harder it is to give up on achieving it.* Lombardi was telling his players that they had already worked too hard to give up—that if they quit, all of their hard work was for nothing.

- You are going to fail occasionally, even when you don't give up and quit. It happens. Never let this be an excuse for failure, but do understand that you don't always win, and you don't always get it right no matter how hard you try. When you give something your best effort and still fail, remember this: now you are better prepared to try again and succeed the next time. A failed attempt is not the same thing as a failure. You don't fail unless you give up—you just haven't succeeded yet. Look at a failed attempt as an opportunity to try again better prepared this time to succeed.

Overcoming Adversity

There will be adversity in your life as you work to build a winning career in electronics. Some of it will be personal, and some will be professional. On the personal side, you might lose a loved one, go through a divorce, or suffer health problems. On the professional side, you might lose your job due to circumstances beyond your control, work for a company that is bought by another company that makes a lot of changes, have more work assigned than you have time to do, or face deadlines that seem impossible to make.

All successful people face adversity in their lives and their work. One of the reasons they are successful is that they learn to face it squarely and overcome it. My favorite example of a person who had to overcome adversity in order to succeed is Franklin Delano Roosevelt, President of the United States during some of the most difficult times our country has ever faced. When Roosevelt was elected president, the United States was going through the Great Depression.

During the Great Depression, unemployment was at its highest level in our country's history, the nation's banking system had crashed causing millions to lose their life savings, small businesses were going out of business on a daily basis, and people were losing their homes because they could no longer pay the mortgage.

This is the situation as it existed when Franklin Delano Roosevelt was elected. In spite of the challenges facing him and our nation, Roosevelt practically exuded confidence and optimism. He used radio broadcasts that came to be know as "fireside chats" to tell Americans that ". . . the only thing we have to fear is fear itself." He assured citizens in a calm and optimistic voice that the economy would turn around. His calm optimism began to give people hope. Roosevelt then began to push a long list of programs through Congress aimed at turning the economy around and getting people back to work.

However, before Americans saw any real improvements in their lives, disaster struck. The Japanese attacked America's military bases at Pearl

Harbor, Hawaii, and the United States was suddenly plunged into World War II. In just one day, Americans went from worrying about jobs and feeding their families to fighting a world war against the Japanese, Germans, and Italians. This was an enormous burden of adversity for a country and its president to have to bear.

Determination and perseverance were called for in facing this new form of adversity, and once again, President Roosevelt provided the leadership needed. Roosevelt applied the same optimism and perseverance to face the adversity of World War II that he had encouraged Americans to apply in facing the Great Depression. His determination and perseverance were incredible by themselves, but what is even more incredible is that he led the country through the war while suffering from polio. President Roosevelt could not even walk—a fact he felt compelled to hide from the world because in those less enlightened times, he did not want Americans or the enemy to view him as weak.

In addition to having to lead our country in a war for which it was unprepared, President Roosevelt had to struggle every minute of every day against the painful effects of a crippling disease. To his eternal credit, Roosevelt persevered against the odds, the pain, and the demands of an office that will wear out a person who is in the best of health.

Whenever in public, Roosevelt went to great lengths to disguise his physical disability, even though it meant supporting himself with heavy steel braces that bit painfully into his frail and paralyzed legs. Roosevelt's physical condition increased the burden of adversity this determined and courageous leader had to face day in and day out as president of the United States. Until he died at his "Little White House" in Warm Springs, Georgia, Roosevelt never lost his courageous spirit of optimism and perseverance in the face of adversity. When you face tough times in your life, remember the example of President Franklin Delano Roosevelt.

When Work Is Difficult—Just Keep Going

One of the reasons only a relatively small number of people enjoy a large measure of success in their careers is the human tendency to quit when times get tough. A fact of life that people who work soon learn is that life on the job is sometimes difficult. Organizations that hire electronics technicians operate in an intensely competitive global environment that includes companies not just from the United States but also from Japan, China, Korea, Indonesia, Malaysia, and other industrialized nations. Global competition means that companies in the United States need their electronics technicians to work harder, smarter, and—on occasion—longer than their counterparts in other industrialized countries.

How To Keep Going
When Work Becomes Difficult

- Look "down the road" to when things will be better.
- Don't let yourself get hung up on the unfairness of the situation—move on.
- Stay positive.

FIGURE 12-1 When you feel like quitting, try these strategies.

In addition to the competition, there will be many other types of challenges as you seek to build a winning career for yourself. It is when work becomes difficult that perseverance becomes especially important. The best advice I can give you for when work becomes a challenge and you want to quit is this: *just keep going.*

Of course, advice is easier to give than it is to take. Consequently, rather than just advise you to persevere when your work becomes difficult, I have provided the following strategies that will help you stick it out during tough times on the job: 1) look beyond the tough times to when times will be better, 2) let go of feelings of unfairness and move on, and 3) no matter what happens, stay positive (see Figure 12-1). These strategies for persevering during tough times are explained in more depth in the following sections.

Look Beyond the Tough Times

Assume that you have been working hard to win a promotion, but the promotion goes to someone else. You would be understandably disappointed and probably a little dispirited if this happened to you. Assume that you have asked for a transfer to a branch of your company that is closer to your home, but the transfer goes to someone else. Again, you would be understandably disappointed. Unfortunately, things like this happen in the workplace. Disappointments are not uncommon among people who work in any profession. The workplace is not always a fair place.

Sometimes you will do your best—you will do everything right—and still not win. I hope this will not happen to you, but it might. If it does, the best approach is to look to the future at a place beyond your immediate disappointment. In other words, don't let yourself get caught up in the unfairness of what has happened. Instead, set your sights farther down the road on winning the next time there is an opportunity for a promotion or a transfer. Then begin immediately working hard and working smart to get it.

A case study in looking beyond the tough times

I know of an electronics technician who experienced a series of set-backs and became so disappointed that he considered changing professions. When he came to me for counseling, I told him that changing professions wasn't the answer. One profession can be just as unfair as another and that during his career there were going to be frustrating times like those he was currently facing. When he asked me what I thought he should do, I told him to just keep going, forget the unfairness of the situation, and focus on a point down the road when he might have another chance to be named a departmental supervisor (this was the goal he had missed three times when he came to see me).

This electronics technician—I will call him Phillip—had applied for a supervisor's position for the first time two years earlier at the time he came to see me. On his first try, he didn't even get an interview. He was disappointed, but looking back later he could see that at the time he was too inexperienced to be a supervisor, and it was probably best that he had not been selected. To his credit, even though he wasn't even interviewed, he continued to work hard and work smart so that he would have a chance the next time.

The next time a supervisory position came available was a year later. Phillip had done well during this period of time earning recognition as "Employee of the Month" twice. He was sure that the supervisor's job was a good as his, and sure enough, he was given an interview. Phillip thought the interview had gone well, but when the anticipated call came from the manager of the electronics division of his company, it was to tell him that another technician had gotten the job.

Phillip was hurt, disappointed, and angry. He told me that his immediate reaction had been to begin looking for another job convinced that he would never get a promotion at his current company. He was angry with the company, not the person who got the job instead of him. This technician, like Phillip, was good at his job and had been recognized for his outstanding performance. In addition, he had more experience than Phillip. When he had settled down enough to look at the situation logically, Phillip could see that the person who won the supervisory position was better qualified, even if by only a little. Consequently, he determined to hang in there and set his sights on the next supervisor's job that opened up.

When a third opportunity finally came up, Phillip told himself, "This one is mine." Unfortunately, he didn't get the job. This time he was the runner up, and, to make matters worse, the person who did get the job was not as qualified as Phillip. Knowing this really ate at Phillip. The rumor circulating throughout Phillip's company was that the person selected for the supervisory position this time got the job because "he knew someone at the top." To Phillip, this was the worst kind of insult. It was one thing to lose out to someone who was more qualified, but it was unacceptable to lose on account of quasi nepotism. Phillip had had it with his company. That's when he came to see me.

As I listened to Phillip, my initial reaction was to tell him to begin quietly looking for a new job as an electronics technician with another company. When companies promote on the basis of who you know rather than on the basis of merit, it is time to find a better situation. However, I knew this company and its management team well. I was sure this quasi-nepotism approach to promotions was not the norm for this company. Consequently, rather than advise Phillip to leave, I advised him to give it one more chance. If he was turned down for a supervisory job one more time, then leaving would certainly be appropriate.

Phillip did not like this advice, but he took it. Then, in less than a month, I got a call from Phillip inviting me to a party celebrating his promotion to supervisor. Predictably, the person who was promoted because of who he knew instead of what he knew had made a mess of the job. His whole department was in turmoil because of his ineptitude. The unqualified supervisor had been given the opportunity to return to another team as a technician or to be fired. Phillip had been called in and asked to take the position as departmental supervisor. His willingness to look beyond his current disappointments to a better place down the road had finally paid off.

Let Go of Your Disappointment and Move On

Have you ever heard the old saying that you should not cry over spilled milk? It means that once something has gone wrong, the worst thing you can do is fret about it and let it distract you from doing what is necessary to move forward. The best way to handle spilled milk is to just clean it up and get another glass. The same is true of handling the disappointments that come your way in the workplace. Sometimes you have to just let go of disappointments and move on. Notice that I didn't say "forget" about the disappointments. I said "let go" of them. If you just forget about the disappointments in your life, you might not learn anything from them. Letting go means assessing what happened and why, asking yourself if there is anything you could and should have done differently, learning from any mistakes you or anyone else involved might have made, and moving on. This approach ensures that when you move on to try again the next time, you will be better prepared to succeed.

Successful people have many things in common and share many common traits. One of these traits is that they refuse to hold themselves back by focusing on the unfairness of life. They understand that life occasionally is going to be unfair, and they will occasionally be the victim of the unfairness. When you are in a leadership position in electronics, do everything you can to make life fair for your direct reports, customers, suppliers, and yourself. However, understand that the things that make life unfair are often beyond your control. Consequently, bad things can happen to good people—understand this and do not let it discourage you.

There will probably be times in your career when you will not get a raise or promotion that you feel you deserve. There will be times when you will have to work late because of someone else's ineptitude. There will be times when a customer will complain about your work when, in fact, the problem is the customer's fault. There might even come a time when someone else will be recognized for work that you did. The workplace is full of situations such as these. They are not right and are not fair, but they are part of life.

When your job or life in general seems unfair to you, make a concerted effort to look beyond the situation. Do not let yourself get caught up in fuming about the unfairness. The reason for this is simple. The fact that a situation is unfair is not going to change the situation. While you are wasting time and energy distressing yourself over the unfairness of a situation, you could be doing what is necessary to make sure the same thing does not happen to you again. Let go of your anger and frustration and move on.

Maintain a Positive Attitude

When things are not going your way, it is easy to give in to your discouragement and let yourself become cynical and negative. Staying positive when the world around you seems negative is a challenge for even the most positive-minded people. In fact, the truth is that most people cannot do it. But learning to maintain a positive attitude in spite of the difficulties of life is one of the characteristics of successful people (see Figure 12-2). It is one of the characteristics that sets them apart from the less successful.

This is why facing adversity with a positive attitude is so important to you in building a winning career in electronics. If you can stay positive and focused on getting the job done when others cannot, you will

---

**How To Stay Positive
in a Negative Situation**

- Avoid negative people.

- Spend time with positive people.

- Reach out to others who are experiencing difficulty and help them.

FIGURE 12-2 Strategies for staying positive in the face of adversity.

succeed when they don't. Falling into a negative attitude is the easiest thing in the world to do. Anybody can have a negative attitude. It is staying positive in the face of adversity that is hard. Doing so requires commitment, effort, and self-discipline. These are all characteristics of successful people.

It's one thing to say that you should maintain a positive attitude, even in the face of adversity. It's quite another to be able to do it. The following strategies are provided to help you stay positive when the world around you seems negative:

- *Avoid people who seem to always be negative.* Have you ever known someone who just seems to be negative all the time? No matter what the situation happens to be, some people will find the negative in it. Such people are known as "pessimists." A pessimist is someone who focuses on the negative. For example, if you comment to a pessimist that, "It sure is nice weather today," he might say, "Maybe, but not for long." Pessimists can put a negative spin on almost any situation or issue. They tend to be grumblers and grouches. Make a point of avoiding such people. It is hard enough to stay positive in the face of adversity without having to overcome the gloomy attitudes of pessimists.

- *Spend your time with positive people.* When I speak of positive people, I am not referring to people who are artificially positive, that is, people who paste on a fake smile and pretend to be happy even when they aren't. Rather, I am talking about people who, when faced with adversity, confront it with a positive attitude and say, "We have a problem. Let's see what we can do to solve it." I am talking about people who make a point of maintaining an optimistic, positive attitude when times are good as well as when they are bad, but especially when they are bad. Most people can be positive when things are going their way. This is no challenge. You need to make a point of spending time around those who can stay positive in the middle of a crisis and when things are not going well.

- *Help someone else who is experiencing hard times in their job or life.* When facing hard times, one of the best things you can do to help yourself is to help someone else who is facing adversity. Looking beyond your problems to those of someone else can actually help you let go and move on. This strategy can have several benefits:
  - It can show that you are not the only person with problems and that there is always someone whose problems are even bigger than yours—you are not alone.
  - The problems others are facing can take your mind off of those you are facing long enough for you to regain a more positive perspective.
  - It can win you a friend or ally who will help you in the future.

Success rarely happens fast. It can take time, and as people who are accustomed to microwave ovens, ATMs, and the instant responses of computers, we do not like to wait. This is why perseverance is so important to building a winning career in electronics. Success cannot be microwaved. It often requires that you persevere in doing a good job over a long period of time. That is one of the realities of the workplace. However, there is also good news. If you are one of the few who is willing to persevere when times get hard, you can also be one of the few who enjoys a highly successful career as an electronics technician.

Review Questions

1. Explain why it is important to make the following commitment when attempting to build a winning career in electronics: "Never quit and never give up."
2. Analyze the true story of Colonel Bud Day presented in this chapter, and list all of the individual examples of perseverance it contains.
3. List and explain several strategies you can use when you feel like you want to quit.
4. Why is it so important to be able to face adversity in your career and life and overcome it?
5. Sometimes when building a career, you have to *just keep going* when the job becomes difficult. Explain some strategies that will help you just keep going.
6. Why is it so important for aspiring electronics technicians to be able to "let go of disappointment and move on."
7. How can you maintain a positive attitude in the face of adversity when building your career in electronics?
8. Taking everything presented in this chapter into account, how would you answer the following question: "Why is perseverance so important to a person trying to build a winning career in electronics or any other field?"

Discussion Questions

1. Discuss with your classmates the following statement from this chapter: "A career is like a baseball season—it consists of more than one game." How does this apply to you as you strive to build a winning career in electronics?
2. Discuss the following situation with your classmates: An occasion you can remember when you really wanted to just quit and give up, but didn't. What were the circumstances surrounding this occasion, and what did you learn from it?

3. Discuss the case of Colonel Bud Day presented in this chapter. Have you ever heard of any other case in which a person persevered through so much adversity? Discuss any such cases you know about.

4. Discuss with classmates any situations you know of in which a person faced adversity and overcame it. What are the details of the situation? How did the person in question overcome the adversity? What did you learn from this person's example?

5. Have you ever had to just keep going when you felt like stopping? Discuss the situation with classmates. What were the details of the situation? How did you make yourself keep going when you wanted to stop? What did you learn from this situation?

6. Have you ever had to let go of disappointment and just move on? Discuss the situation with your classmates. What were the details of the situation? How were you able to let go of the disappointment and move on? What did you learn from this situation?

7. Do you know someone who is able to maintain a positive attitude in spite of difficulties in his or her life? Discuss this person's situation with your classmates. How does this person's positive attitude make you feel toward him or her?

Application Assignments

1. Conduct the research necessary to identify someone who refused to quit, who would not give up, and as a result ultimately succeeded. Write a brief description of the situation, or make an oral presentation to your class in which you explain it.

2. Conduct the research necessary to identify someone who had to overcome great adversity in order to succeed. Write a brief description of the situation, or make an oral report to your class in which you explain it.

3. Identify someone who works in electronics, and ask this person to tell you about instances in which it was necessary to persevere through adversity to get the job done. Ask for specific examples you can share with your classmates.

APPENDIX

ADVANCED SUCCESS SKILLS

The material in this appendix consists of checklists and guides for developing advanced success skills—skills that will help electronics technicians who climb the career ladder and become supervisors and managers.

Integrity Checkup

Electronics technicians who become supervisors and managers will be expected to set a consistent example of honesty and integrity. The following list of questions will help you determine if there are areas in which you need to improve concerning integrity. The desired answer is "A" for "Always" for each item. If you cannot give this answer for a given item, work on improving in that area until you can.

A = Always
U = Usually
O = Occasionally
N = Never

_____ 1. Everyone I work with can accept my word as my bond.
_____ 2. When I am wrong, I am willing to admit it and take responsibility.

_____   3. I tell the truth even when it hurts to do so.
_____   4. Everyone I work with can trust me to keep promises I make.
_____   5. My standards of conduct are the same every day.
_____   6. I am willing to put the needs of others before my own.
_____   7. I am willing to be held accountable for my actions.
_____   8. I understand that I have to re-earn the trust of others every day.
_____   9. I do not make excuses when I make a mistake.
_____ 10. I will not lie even when it appears to be in my self interest to do so.

Listening Checklist

Listening is a critical skill for electronics technicians, and the higher up the career ladder you go, the more important it becomes. The following list of listening strategies will serve you well when you reach the point where you receive complaints from direct reports, customers, and suppliers:

1. Remove all distractions so that you will not be interrupted.
2. Use a few minutes of small talk to put the individual at ease.
3. Look directly at the person who is talking to you so that he or she knows you are paying attention.
4. Concentrate on what the speaker is saying—don't let your mind wander.
5. Watch for nonverbal cues. Sometimes body language and tone of voice will say as much as the words.
6. Be patient and wait. Don't rush in and finish sentences for the speaker.
7. After the person has stated his or her case, ask any questions that might be necessary to clear up inconsistencies or to elicit additional information.
8. When you think you fully understand what has been said, paraphrase and repeat it back to the speaker for confirmation.
9. No matter what is said or how it is said—control your emotions. Remember, if you lose your temper, you lose period.

Negotiating Checklist

Once you advance to the level of supervisor or manager, you might be asked to join your organization's negotiating team for negotiating labor, supplier, or customer contracts. When this happens, the following checklists will help.

Preparing to Negotiate

- Decide what your side wants out of the negotiations.
- Decide what the other side is likely to want out of the negotiations.
- Decide what is at risk in the negotiations (for example, what do we lose if the negotiations fail?).
- How much do we know about the other side? How can we learn more?
- Does our side have any "hot buttons" in the negotiations?
- Does the other side have any "hot buttons" in the negotiations?
- What don't we know about the other side that we really need to know before the negotiations? How can we learn what we need to know?
- Are there factors that could affect the outcome of the negotiations that we have no control over? If so, what are those factors? How can we mitigate the uncontrollable factors?
- Are there any factors that could affect the outcome of the negotiations that the other side has no control over? If so, what are those factors? How will the other side attempt to mitigate those factors?
- What is our bottom line in the negotiations—the point where we walk away?
- What is the other side's bottom line in the negotiations—the point where they walk away?

Conducting Negotiations

- Negotiate in stages: 1) Lay the foundation by convincing the other side of the value of the potential contract/agreement—in their terms; 2) Discuss the details of the potential contract—but not price; and 3) Close the deal by agreeing to price and dates.
- Select the time and place of the negotiation carefully and to your advantage.
- Project a professional image (for example, make sure everyone on your team is knowledgeable, well-informed, properly dressed, and so on).
- Create early momentum by putting several "easy-to-agree-on" items up front in the negotiations.
- Think critically about everything that is said during the negotiations.
- Listen to what is said as well as what is not said (for example, does the other side appear to be holding information back?).
- Put yourself in the shoes of the other side, and keep their hopes and needs in mind.

- Be patient—don't try to rush the negotiations.
- Don't let the other side get to you by making personal comments—ignore such tactics.
- Don't paint yourself into a corner. Always leave yourself some room to maneuver.

Final Strategy

Remember that the best negotiation is one that paves the way for additional business in the future. You don't win by being sneaky or by out-witting the other side so that they are left with nothing. If this happens, your organization will pay in the future with lost opportunities. The best negotiation is a *partnership* that leads to a win-win conclusion.

Public Speaking Tips

As you climb up the career ladder, the inevitable will eventually happen. You will be asked to make a presentation to your company's higher management team or to an important customer. When this happens, it is important to make the right impression as well as to make a good presentation. The following public speaking tips will help.

Prepare Yourself and Your Materials

Prepare yourself by doing the necessary research, gathering together all necessary materials, and planning what you intend to say. Develop the outline of key points you want to make, and practice the presentation several times before you have to give it. Make sure that you time your practice sessions so that you can easily make the presentation in the allotted time.

- Keep your materials simple.
- Use bullet points and key statements on all visual materials (Power-Point slides, handouts, and so on).
- Think BIG. Prepare all visual material as if the audience will have eyesight problems (many will).
- Build in redundancy. Don't rely on just one approach. Make sure you can still make the presentation if something goes wrong with your computer, LCD, or other technologies. Handouts are a good way to back up electronic visuals.
- Put something in their hands. Always give your audience something that summarizes your presentation (copies of your PowerPoint slides, brochure, and so on), and bring more copies than you think you will need.

Prepare the Facility

Never take for granted that the facility (conference room, auditorium, and so on) where you will make your presentation will adequately accommodate your needs.

- Conduct a reconnaissance mission ahead of time. Visit the facility if possible a day or two before the presentation. If this is not possible, arrive early enough that you will have time to make adjustments if necessary.
- Determine how to operate the lighting.
- Locate the microphone if you plan to use one, and make sure it works.
- Locate the podium if you plan to use one, and make sure it is where you need it to be.
- Locate electrical outlets, and make sure your technology can be properly plugged in.
- Locate flipcharts and marker boards if you plan to use them, and make sure there are adequate pages left on the flipchart and that the markers are available and not dried out.
- Determine if the temperature in the room can be controlled and how to do this. The room may become uncomfortably stuffy when too many people are crowded into it.

Make the Presentation

Once you have properly prepared, all that remains is to make the presentation. The following tips will help make the presentation go well:

- Maintain eye contact with your audience whether it is 2 people or 200 people.
- Show some enthusiasm for your topic. They won't get excited about what you have to say if you don't.
- Be yourself. Don't adopt any made-up gestures or voice tones. Your presentation should be a conversation between you and the audience regardless of the size of the crowd.
- Don't say what you don't know. If someone asks a question you cannot answer, don't attempt to act like you know. Instead, just say "I don't want to give you an inaccurate answer—let me get back to you today on that question."
- Allow the audience to ask questions and listen carefully to what is asked. Do not assume you know where the questioner is going, interrupt, jump ahead, and answer. Wait and listen carefully.
- Don't rush. If you are nervous about making the presentation, you will tend to talk faster—probably a subconscious attempt to get it over with. Slow down and take your time.

Organizational Culture

An organization might be as small as a team or as large as a corporation or any size in between. When you climb the career ladder to the point that you supervise a team, department, or any other size of organization, the issue of organizational culture will become important to you. An organization's *culture* is the collection of tacit assumptions, unwritten rules, and usual ways of doing things in the organization. You will want yours to be a high-performance organization in which employees give their best effort every day and continually improve their performance. One of the first steps in establishing a high-performance organizational culture is to assess the existing culture. This will show you where improvements need to be made. As you observe your direct reports, ask yourself the following questions as a way to assess the status of the culture in your organization:

- Do all of my direct reports understand the organization's mission?
- Do all of my direct reports understand their roles in helping the organization accomplish its mission?
- Are all of my direct reports committed to peak performance and continual improvement?
- Do we have open, effective communication in our organization?
- Do we mutually support each other in our organization?
- Are all of my direct reports committed to customer satisfaction on every contract or with every transaction?
- Are all of my direct reports willing to participate in a positive way in the decision-making process?
- Do all of my direct reports accept responsibility for their jobs and performance?
- Do all of my direct reports expect to be held accountable for their performance on the job?
- Do all of my direct reports adhere to high ethical standards?
- Do all of my direct reports actively participate in the training that is provided for them?
- Can all of my direct reports disagree without being disagreeable?

Trust-Building Checklist

As you climb the career ladder and begin to have other electronics technicians report to you as their team leader or supervisor, remember that they will not follow you if they don't trust you. To build trust, apply the following strategies:

- Communicate openly and honestly with your direct reports at all times. If the news is good, tell them. If the news is bad, tell them. Always be tactful, but keep your direct reports fully informed and up to date.
- Work on continually improving interpersonal relationships with your direct reports. You are not their "buddy," but they do need to be able to bring their problems, recommendations, and complaints to you and know that you won't "shoot the messenger."
- Be fair and impartial in *refereeing* conflict between your direct reports. Base any conflict-resolution decisions you make on what is best for the organization in terms of accomplishing its mission.
- Involve your direct reports when making decisions that will affect them. Never surprise your personnel with a decision. Give them a voice and listen to what they have to say before finalizing your decision. They might see something in the issue that you don't.
- Promote teamwork by recognizing and rewarding direct reports who put the team's mission ahead of their own agendas.
- When things go wrong, step forward to take the blame for your team. When things go right, pass along the credit to your direct reports.

Occupational Safety

As you climb the career ladder and begin to supervise other personnel, occupational safety will become an important concern. Workplace accidents and injuries cost employers millions of dollars every year, making it even more difficult for them to compete. The best work environment for promoting peak performance and continual improvement is a safe and healthy environment. The following strategies will help you maintain such an environment for your direct reports.

Preventing Cumulative Trauma Disorders (CTDs)

Cumulative trauma disorders are common among electronics technicians. They are caused by repetitive motion and affect the body's soft tissues (tendons, ligaments, and so on). To prevent CTDs in your organization, apply the following strategies:

- Teach your direct reports the warning signs of CTDs (weakness in the hands and forearms, numbness, tingling, heaviness, stiffness, lack of control of the hands, and tenderness to the touch).

- Teach your direct reports to stretch before beginning work. If they perform repetitive motions as part of their jobs, they need to warm up and stretch just like athletes.
- Teach your direct reports to start slowly and increase their pace over time.
- Avoid the use of wrist splints. Teach your direct reports to keep their wrists in a neutral position without the help of splints. Splints can cause the muscles they are protecting to atrophy over time.
- Use hand tools wisely (see the next checklist).

Safe Use of Hand Tools by Electronics Technicians

Electronics technicians use a variety of different hand tools to do their jobs. If not used properly and carefully, cumulative trauma disorders (CTDs) can result. To help prevents CTDs, you should 1) reduce repetitive motion, 2) reduce the amount of force the technician must apply, and 3) minimize awkward postures.

Reduce repetition
- Limit overtime.
- Change the process.
- Provide mechanical assist devices.
- Require periodic breaks.
- Encourage stretching exercises.
- Automate if possible.
- Rotate employees in high-repetition jobs.
- Distribute high-repetition work among more employees.

Reduce the amount of force the technician must apply
- Use power tools whenever possible.
- Use power grips instead of pinch grips.
- Spread the force over the widest possible area.
- Eliminate slippery hard, and sharp gripping surfaces.
- Use mechanical assistance devices to eliminate the pinch grip.

Minimize uncomfortable and awkward positions
- Keep the wrist in a neutral position.
- Keep elbows close to the body.
- Avoid any work that requires reaching overhead.
- Minimize work that requires forearm rotation.

Preventing Back Injuries

Improper lifting techniques cause one of the most common types of injuries to electronics technicians—back injuries. The following lifting techniques will help your direct reports prevent back injuries in your organization.

Plan before lifting

- Check the load first. Can you lift it easily or is it too heavy or awkward?
- Check the route you will use to carry the load, and remove any obstacles or mitigate any slippery surfaces.
- Get help if you cannot easily lift and carry the load yourself.

Lift with your legs

- Keep your back straight and bend at the knees.
- Position your feet close to the object to be lifted.
- Center your body over the object to be lifted.
- Lift straight up, keeping your back straight and using your legs. Do not jerk up—the smoother the motion the better.
- Do not twist your torso while lifting or carrying the object.
- Set the object down slowly and carefully. Again, keep the back straight and bend the legs. Don't let go of the object until it is down.

Push objects rather than pulling them

- Never pull an object. Pushing the object instead will put much less stress on your back.
- Put rollers under the object if possible to make it easier to push.

Preventing Electricity-Related Injuries

Electricity- and static-electricity-related injuries are common among electronics technicians. The following strategies will help your direct reports prevent these types of injuries:

- De-energize an electrically powered system before working on it. Remember that capacitors can store current after the power has been shut off.
- Allow only properly trained technicians to work on electrical/electronic systems.
- Do not wear conductive material such as jewelry when working on electrical/electronic systems.
- Periodically inspect insulation on wires.

- If working on a hot system, wear protective gear, and use the buddy system.
- Do not use a fuse with greater capacity than prescribed for the circuit.
- Verify circuit voltages before beginning work on a system.
- Never use water to put out an electrical fire.
- Check the entire length of an electrical cord for wear and bare spots before using it.
- Use only explosion-proof devices and nonsparking switches around flammable liquids.
- Enclose insulated conductors in protective areas.
- Discharge capacitors before working on equipment.
- Use fuses and circuit breakers for protection against excess current.
- Provide lightning protection on all structures.
- Train electronics technicians in first aid and CPR.

INDEX